Plastics in Packing

Plastics in Packing

Edited by:
CRISTINA GARCÍA JAIME

Delve
PUBLISHING
www.delvepublishing.com

Plastics in Packing

Cristina García Jaime

Delve Publishing
2010 Winston Park Drive,
2nd Floor
Oakville, ON L6H 5R7
Canada
www.delvepublishing.com
Tel: 001-289-291-7705
001-905-616-2116
Fax: 001-289-291-7601
Email: orders@arclereducation.com

ISBN: 978-1-77361-286-7 (Hardcover)

ABOUT THE EDITOR

Cristina Garcia Jaime is a Spanish food specialist who studied Food Quality and Technology in the University of Granada and Human Nutrition and Dietetics in Pablo de Olavide University (Seville). She has worked in different food areas: scientific, industry and clinic. Currently, she has founded a company established in Europe and called "Foodise" dedicated to help small business and companies to manage all issues related to food.

TABLE OF CONTENTS

LIST OF FIGURES

LIST OF TABLES

PREFACE

Under the current circumstance of the world economic downturn, the use of plastic products in packaging industry has maintained an annual growth rate of over 10%. For instance, in 2012 a country as China reached it self more than1.5 trillion RMB in the packaging industry as the gross output. It is also worth noting that food industry and food technology have also maintained a rapid global growth with different new materials, new technologies, and new products emerging every day.

This book has been prepared to explore the relevance of three major themes: plastics, packaging industry, and the environmental threats posed by plastic materials. Being relevanta technical understanding for the packaging industry. It is vital a technical understanding of all the protection requirements of goods and at the same time,it is important to ensure that plastics do not break with a safe environment.

The growing economy, the rapidly growing consumption rates, and the production patterns of various commodities have resulted into rapid growth in the generation of plastic waste materials in the world. Statistically, the total world consumption of plastic in packaging industry has gone up from 5 million tons in 1950s to about 100 million tons currently. This translates to about 20 times more plastic production after more than 50 years. This means that, on one hand more resources are being used in order to meet the current demand of plastic materials in the packaging industry, but on the other hand more plastic materials are being pumped into the environment as waste. Plastic waste materials are the second most municipal solid wastes after food waste materials. Even the smallest cities in the world are facing the problem of plastic waste materials due to the increased use of plastic products in the packaging of goods.

The purpose of this book is to offer important insights and reference different aspects of plastic in packaging industry to those involved. This book has been prepared to incorporate all the relevant information that can be used by students that are taking courses in different fields (such as plastics process manufacturing, production and packaging management, environmental planning and management, urban and regional planning among others), teachers, environmental management staffs, and even various governmental staffs involved in environmental management sectors.

This book comprises ten chapters. Chapter one introduces various aspects in plastics in food packaging. This includes things such as identifying various types of plastic used in packaging industry, brief introduction of various issues of plastic products in packaging industry, how plastic usage in packaging industry has being trending, and a brief understanding on the policy instruments that can be used to regulate and management plastic products in the society among other things. Chapter two analyses the use of plastic packaging materials in the food industry. Chapter three discusses on production of packaging plastic products. Chapter four explains the situation of plastic waste materials in the environment, particularly on marine ecosystem. Some of the key sub-topics discussed under this chapter include characteristics of plastic debris found in marine water bodies, how those debris enter into marine waters, how they affect the marine ecosystem, and solutions to these plastic debris in marine ecosystems. Chapter five deal with various challenges of plastic packaging materials on the dry land. Some of the key sub-topics discussed under this chapter include generation of plastic waste, causes for plastic waste, and management of plastic waste materials.

Chapter six explores key challenges of plastic materials on human health. Chapter seven explores key actors (both formal and informal) in the management of plastic waste materials used in packaging industry. Chapter eight discusses on key challenges and opportunities on reusing and recycling plastic waste materials used in packaging sector. Chapter nine discusses on sustainable plastic waste management through energy recovery technology. Chapter ten explores on legal framework with regards to plastic waste management. Some of key sub-topics discussed under this chapter include environmental laws on plastic waste pollution, current state of the environment in the world with regards to plastic waste pollution, and implementation of various laws and policies used to reduce plastic consumption.

Editor

1 INTRODUCTION TO PLASTICS IN PACKAGING INDUSTRY

INTRODUCTION

The word "plastic" suggests a single material but there are several hundred different plastic polymers. In practice, these polymers can actually be enhanced through the use of inherent properties of a wide range of additives. In other words, these polymers can be engineered into a specific design in order to come up with a unique plastic product for a specific packaging purpose. There are different types of plastic products used in packaging industry, but for the sake of this book, the focus will be towards the packaging plastic bags and containers (both Low-density polyethylene (LDPE) and high-density polyethylene (HDPE)). Studies have shown that plastic packaging products are a big threat to the climate change and the fossil fuel. This is because they are usually made from petroleum products. For instance, about 15 billion barrels of oil is required in U.S.A. only to manufacture about 100 billion of plastic bags every year.

Source: http://www.wastewiseproductsinc.com/blog/sustainability/can-edible-packaging-reduce-the-amount-of-plastic-used-by-the-food-industry/

Figure 1.1: Illustrating key plastic products used in food packaging.

According to American Chemistry Society (2005) viewpoints, plastic bags are widely used by the retailers to package goods at the shopping malls, on the streets and at the market centres. This is because plastic bags have proven to be durable, versatile, and lightweight. Due to over use, most plastic bags have become waste after they have served the purpose of carrying goods from the market centres, shops, and streets. Most of the time, people have practically no use for plastic products especially for the thin plastic bags after they are used.

Some studies have shown that from all the municipal solid wastes collected daily, plastic waste materials have gained popularity because plastic is a functional material, is also lightweight, strong, cheap and can be used to transport food and goods in a hygienic way. It has also been gaining popularity for its grave impacts to the ecosystem. According to World Health Organization (2005) reports, it is estimated that between 500 billion and 1 trillion plastic bags are used globally each year. Much of this waste ends up in the waste stream without being recovered. The mounting concerns about environmental degradation and resource scarcity are trends that have become a manifestation of linear production and consumption models of taking, making and disposing. There is a strong focus placed worldwide on the environmental implications of plastic bags.

Solid waste management studies are an important contribution that links the built environment to the natural environment. Some studies carried outaddress plastic bags waste management and the need to integrate innovation into existing systems for better relationships. However, there is little mention of the influence that human behaviour has on plastic management.

A BRIEF HISTORY ABOUT THE INVENTION OF PLASTIC AND THE DANGER THEY CAUSE

Before the invention of plastics, wood, stone, metals, clays, animal skin etc. were among the many materials man used in making his clothing in the earliest times. However, in recent times, plastics have become a necessary component of man's life ranging from his private to his professional life. Right from the break of day till sunset, man seems to depend almost entirely on plastics and its related materials for everyday survival. One of the many experiments that led to the invention of plastics took place around 1835 when Regnault, a renowned French chemist created a chemical called Vinyl chloride which he turned into white powder. This white powder created some years ago has been modified and known today as PVC.

Plastics were unknown until Alexander Parkes in 1855 publicly displayed it at the Great International Exhibition in London. Plastics, then called Parkesine, named after Parkes, were made by mixing Camphor (chemical used in mothballs) and nitrocellulose (chemical used in many modern lacquers for motor car bodies).

Major breakthroughs in the effort to manufacture plastic were recorded after Parkes in 1855. However, the acceptance of plastics as a useful material for use both at home and at work and the ever increasing demand for more goods, more chemicals, more steel and power led to the wider acceptance and growth of the plastic products industry. In the 19th century, plastics became wide spread and more researches were conducted into finding newer versions of existing plastics for improved uses.

Since the mid-19th century, there has been a significant push in plastics technology to a point where the familiar question, "Paper or Plastic?" is no longer heard. What seemed like such a miraculous invention has turned into the dominant option and an environmental nightmare.

Today, the UNEP (2009) defines plastics "as a polymer, a very large molecule made up of smaller units called monomers which are joined together in a chain by a process called polymerization. The polymers generally contain carbon and hydrogen with other elements such as oxygen, nitrogen, chlorine or fluorine." Researchers have further explained that plastics are polymers and a polymer is an organic macromolecule composed of several hundreds and thousands of repeating segment linked together in chain-like form.

These polymers come together to form a useful versatile material (plastic) which has many uses and has been used extensively in the manufacturing industry in the last three decades. Plastics are durable, cheap, lightweight, strong, inert, which makes them more appropriate for the manufacture of a range of products. These same qualities make them a serious social hazard as form non bio-degradable products that rather fragment into smaller pieces known as plastic pellets.

While plastics may solve many issues for human's daily lives, it causes many more problems into the environment, whether on land or in waterways. Just in 2010, an estimated of 8 million tonnes of plastic found its way into the ocean via coastal countries. However, the anti-plastic movement has gained speed in the last decade and a half as it has entered developed countries. Through use of different policy tools, such as taxes or bans, many countries have made an effort to reduce the consumption of plastic products used as packaging or carrier materials.

The pollution that plastics cause begins before the production process. The extraction of oil and natural gas used in the production of plastics causes greenhouse gas emissions such as carbon dioxide and hydrogen sulphur. However, the damages that plastics cause do not end with their production and use of fossil fuels such as oil and natural gas. When discarded, these plastic products remain in the environment harming animals when they confuse them with food. Plastic bags are very prone to travel long distances due to their light weight parachute design and thus pick up chemicals and contaminants along the way that could poison waterways and wildlife. According to UNEP (2005) report, it is estimated that half of the earth's sea turtles have ingested plastics, while around 90% of seabirds have consumed plastic debris. The ingestion of plastic causes major blockage issues within the intestines, causing failures to the digestive system resulting in death to many of these animals.

In March 2011, an article in the Daily Mail reported an incident where a sea turtle was found with more than a square foot of varying colours and shapes of plastic debris in its stomach. Sea turtles confused plastics with jellyfish when looking for a meal. Plastic bags take the shape of jellyfish while floating in water. Humans may be able to tell the difference but these sea turtles cannot. Many of the plastics found were sharp which give rise to the possibility of puncturing vital organs in addition to the already present danger to the digestive system. There are dozens of pictures of turtles entangled in plastics as well. In addition, many pictures show dead seabirds with open stomachs featuring plastics including bottle lids, lighters and dozens of plastic fragments. The anti-plastic movement could not be more important since many of the world's wildlife are threatened each day by the existence of plastics within their environment.

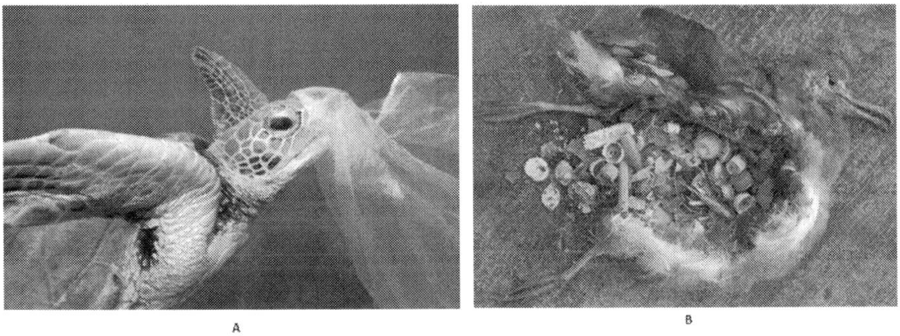

A B

Sources: http://www.iflscience.com/plants-and-animals/seabirds-are-eating-plastic-litter-our-oceans-not-only-where-you-d-expect/https://twitter.com/justinrfield/status/836149035111075841

Figure 1.2: (A) Illustrating sea-turtles mistakenly consuming plastic bags thinking its jerry fish (B) Illustrating dead seabirds with open stomachs featuring plastics including bottle lids, lighters and dozens of plastic fragments.

HOW TO IDENTIFY DIFFERENT TYPES OF PLASTIC

Since the invention of plastic in 18th century, many products made of plastic are emerged. But it is in the food sector where the use of plastic has a dominant role. Plastics are widely used to package different commodities across the world. The following is a breakdown of different types of plastic used in today's society.

Polyethylene Terephthalate (PET)

Source: http://eshrm.com/gallery/#

Figure 1.3: Polyethylene Terephthalate (PET).

Generally, these plastics are usually tough and clear. Sometimes they can be used as fibers. Some of the key uses of these plastics include packaging of mineral water and soft drinks, textile fibers, and are sometimes used for filling pillows and sleeping bags. When recycled, they are used for packaging carpet fibres, detergent bottles and fleecy jackets.

High-Density Polyethylene (HDPE)

These are the most common plastics. They are either coloured or colourless. Some of the areas where these plastics are used include shopping bags, cream and milk bottles, shampoo bottles, freezer plastic bags, and milk crates. When recycled they are used to make agricultural pipe pallets, compost bins, kerbside recycling boxes, detergent bottle crates, and other mobile rubbish bins.

Source: http://www.recyclecare.com.au/en/Grade_HDPE.html

Figure 1.4: High-density polyethylene (HDPE).

Low-Density Polyethylene (LDPE)

These plastics can be categorized as the soft and flexible plastics. They are mainly used to make rubbish bins, lids for the ice cream containers, black plastic sheets, and various bin plastic bags. When recycled they are used to make builders films, packaging nurseries, industry, and other types of plastic bags.

Source:http://www.recyclecare.com.au/en/Grade_LDPE.html

Figure 1.5: Low-Density Polyethylene (LDPE).

Un-plasticized Polyvinyl Chloride (UPVC)

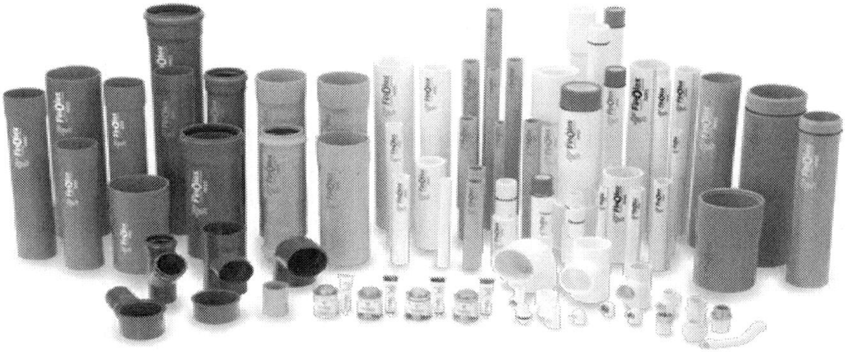

Source:http://www.recyclecare.com.au/en/Grade_V.html

Figure 1.6: Un-plasticized Polyvinyl Chloride (UPVC).

These plastics are mainly rigid and hard and they may be clear or coloured. They are mainly used to manufacture bottle for juice packaging, various plumbing pipes and fittings, and blister packs. When recycled, they can be used to make various bottles for detergent packaging, plumbing fittings and pipes, and tiles among other products.

Plasticized Polyvinyl Chloride (PPVC)

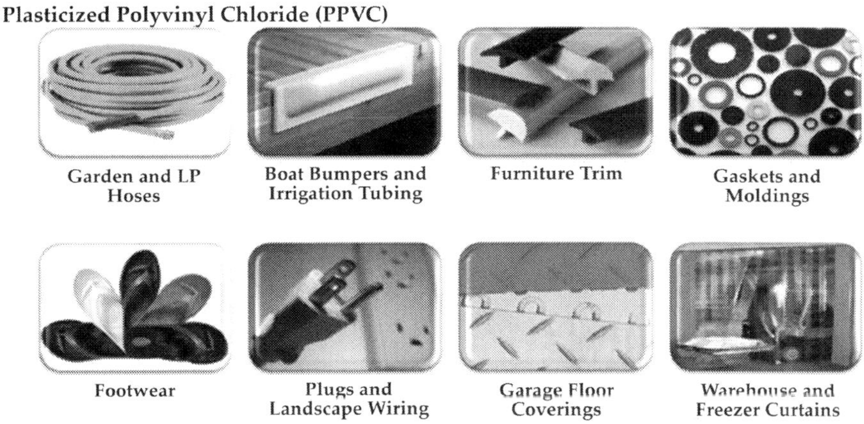

Source:https://buildingmaterials.com.my/materials/polyvinyl-chloride

Figure 1.7: Plasticized Polyvinyl Chloride (PPVC).

These plastics are usually clear and elastic in nature. They are mainly used to make the garden hose pipes, tubing, plastic bags for blood packaging, and shoe-soles. When recycled, they can be used to manufacture inner core for hose pipes and also can be used to floor various industrial houses.

Polypropylene (PP)

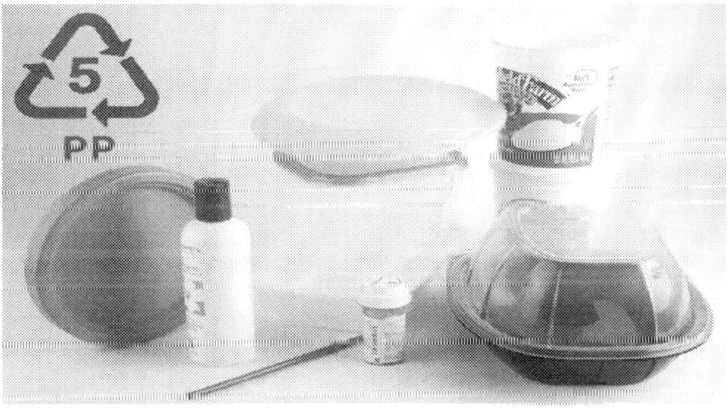

Source:http://www.recyclecare.com.au/en/Grade_PP.html

Figure 1.8: Polypropylene (PP).

Despite their hard structure, they are usually flexible in nature. They are used for different purposes such as manufacturing containers for ice cream packaging, bags for potato crisp, lunch boxes, and drinking straws. When recycled they can be used to make compost bins, wormeries, and kerbside boxes.

Polystyrene (PS)

Source:http://www.recyclecare.com.au/en/Grade_PS.html

Figure 1.9: Polystyrene (PS).

These plastics are generally brittle and rigid. In terms of appearance they are mainly clear and glassy. They are also lightweight and foamed. They are used to manufacture plastic products for yogurt packaging, plastic cutlery, cups for hot drinks, and containers for takeaway food packaging. When recycled, they can be used to make pegs, hangers, various accessories used in offices, rulers, spools, DVDs boxes among others.

The other category includes all the other plastics used in packaging and storage. These include the nylon and the acrylic plastics. Mainly these plastics are not recyclable.

A BRIEF DISCUSSION ON PLASTIC PRODUCTS ISSUES

Billions of plastic bags have been produced since their invention. This depletes our earth's non-renewable resources, such as chemicals, energy and petroleum-based products. For instance, studies have shown that United States alone uses over 100 billion bags annually which is equivalent to throwing away over 12 million barrels of oil per year. Marine debris, about 8 million tons of plastic wastes finds its way into the world's oceans each year. A successful management of the problem requires a comprehensive understanding of both marine debris and human behaviour.

Plastics waste concern has been expressed by many stakeholders, various government organizations, environmental Non-Governmental Organizations (NGOs) and the public at large. During a visit to the US, Prince Charles told leading delegates to an environmental conference that the solution to problems caused by a throw-away society was to move towards a circular economy, where materials are recovered, recycled and reused instead of created, used and then thrown away.

The greatest challenge to the near-future environment is in how people can grasp the full potential within the environment to create a new relationship with the natural world, so that they are not merely depleting and polluting the environment with toxic waste, but purifying and enriching it. Notions of environmental justice are now a prominent part of contemporary discussions of the meaning and practice of sustainable development and take the moral concerns further.

Scientifically, plastic packaging products are usually derived from crude oil and people are consuming these natural resources at a rate 50 per cent faster than the earth can produce them. Electricity and transport are still based mainly on the burning of fossil fuels, a process that is emitting carbon dioxide that is contributing to climate change. These observations indicate that the current use of plastics is not sustainable.

One significant environment that is usually affected by plastic bag waste pollution is the marine ecosystems. According to a claim by the UN, tests of ocean water have found in some areas, the ratio is a ten to one ratio (10:1) of plastic to plankton. The researches have shown that by nature, plastic attracts oily pollutants and nertles, which resemble fish eggs, a common food for animals in the marine ecosystem. Many ocean fish consume nertles coated in toxic pollutants and humans then consume the toxins (which are potentially harmful to their health) when they eat the fish. It has been explained that the quantity of plastic waste that are left unchecked, 17.5 million tons a year, could be entering the oceans by 2025. Cumulatively, that is 155 million tons between now (8 million tons of plastic waste per year) and then. There is a lack of data on the accumulation of plastic debris in natural terrestrial and freshwater habitats; however, it seems that the quantity of debris in the environment will continue to increase unless people change their practices.

Source:https://www.timesofmalta.com/articles/view/20170820/business-news/Plastic-packaging-forms-the-coffin-of-marine-life.656049

Figure 1.10: Packaging plastic bags in natural terrestrial and freshwater habitats.

Globally, the problem of plastic waste pollution is not only a problem in various developing countries, but is also an issue in various already developed countries. The landscapes in various developing countries such as Africa and some parts of the Asian regions are described as having "filmy bags float through the air, swathe the natural foliage, fill ditches and hang from branches."

Source:http://theconversation.com/paper-or-plastic-how-disposable-bag-bans-fees-and-taxes-affect-consumer-behavior-48858

Figure 1.11: Illustrating landfills completely filled with packaging plastic bags.

There is the need to address the perceptions and behaviours of different people around the world with regard to littering plastics material in the environment. This can be achieved through education programs and campaigns. Plastics have become a menace in most of the public places, especially in large cities and towns across the world. This has put more pressure to various solid waste managers. According to Word Bank (2005) suggestions, "more attention and resources need to be devoted to increase the participation and awareness of consumers about the environmental consequences of litter."

An increasing population and growing demand for consumer goods increase the per capita rate of waste generation in different cities around the world. The UN projection is for world urban population to become majority, 66 per cent, in 2050.

Plastics production and transportation not only does it consume energy, but also contributes greatly towards depleting the natural resources. According to UNEP (2005) report, it has been shown that plastics are key source of global warming emissions. Production of plastics emits toxic chemicals and fumes such as methane, ethane, and ketones. Research surveys have indicated that every 520 single-use High Density Polythene

(HDPE) bagsresult in 3.12 kg of materials consumed, 210 mega joules (MJ) of energy used, and 6.08 kg of carbon dioxide greenhouse gas emitted. Similarly, the same number of single-use Low-Density Polythene (LDPE) bags results in 11.77 kg of materials consumed, 957 MJ of energy used, and 29.8 kg carbon dioxide greenhouse gas emitted. Putting this into consideration, to supply the US with plastic bags use (90-100 billion), it takes 12 million barrels of oil to produce.

There have been significant threats on human health due to plastic waste. Most plastics are relatively inert biologically, and problems with their use largely result from the presence of trace amounts of non-plastic components e.g., monomers and plasticizers. This has led to restrictions on the use of some plastics for food applications. For example, the use of polyacrylonitrile for beverage bottles was banned because traces of its monomer, acrylonitrile, were a possible carcinogen. There has been concern about endocrine disruption from phthalate-containing plasticizers used in manufacturing of plastics (e.g. PVC). Safer plastics include PET usually for soft drinks, water bottles, ketchup and salad dressing. Plastics to avoid include PVC used to make toys, especially because children put everything in their mouths. They leach as they age (mostly into fatty foods). PVC is a public health hazard because it contains dioxin and other persistent organic pollutants that can leach into the surrounding environment over time. There is also a link between phthalates and early onset of puberty. There is also suggestion that phthalates are suspect as human cancer-causing agents, could damage the liver and kidneys, and might damage the development of reproductive organs.

TRENDS OF PLASTICS USE

Studies have shown that most people around the world use plastics for different purposes. Majority of people use plastics for shopping, for selling items, and also for storage. The use of plastic has received much attention in recent years and a lot of work has been done in order to understand the significance of the plastic to retailers and consumers. For instance, the need for durable shopping bags led to the innovation of plastic bags. According to the Innovation theory, researchers identified that the recognition of a need or problem as the first step towards any innovation. In other words it means that every innovation must solve a particular problem in the society.

There are different reasons why most people prefer using plastic for their packaging activities. It is believed that most people prefer plastic because they are easily available in the market. Plastic is also lighter compared to other packaging materials. It is also important to note that plastics are usually cheaper to acquire. Most often they are provided for free especially when doing shopping in the market after every purchase.

In recent times, plastic packaging products have received the attention of the media because of their sanitation problems. There have been efforts by both the government and the private sector towards environmental sanitation. Plastic bags particularly have been singled out as the number one cause of environmental challenge. The proportion of plastic waste found in the solid waste composition has equally increased over the years. An estimation of the proportion of solid waste generation from different cities across the world in 2008 revealed that plastics accounted for 15%. This is an indication that the use of plastics has increased in recent years. Studies have also revealed that the demand for plastic use over the last few decades has immensely increased due to an increment in urbanization and rapid population growth rate. Emerging business and trading activities in different cities around the world have also contributed to the increase in the use of plastics.

The use of plastics has also increased in recent years because of the earlier responses by consumers. There is a lack of alternative and therefore it seems to be the only means through which goods bought from the market are packaged and carried home. In other words, plastic usage is a well accepted practice these days.

Interestingly, perception has also contributed in terms of why most people use plastic products in their daily activities. For instance, studies have revealed that everything bought these days is packaged in plastic. They are used to sell almost everything in the market. Usually, these plasticshave attractive designs that make them enticing. Black plastic bags which are also the ones normally used in the market are known for their ability to conceal goods from public view.

It can be concluded that plastics serve varying purposes for consumers. Plastics are also widely used because there are no alternatives, they are low-priced however very easy to handle and moreover people get them for free after every purchase of goods at the market. Plastics have indeed been widely commercialized and hence have penetrated into the society to be fully adopted by consumers and retailers in the market. For instance, plastic bags gained quick acceptance and their diffusion has been fast and rapid.

According to American Chemistry Society (2005), it has been found out most people around the world prefer plastic (bags and containers) because of their durability, lightweight, and versatility. It has been revealed that the uniqueness of the plastics has made them acceptable and widely used in almost every society in the world. In the US alone, 10 billion of plastics are used each year, costing retailers $4 billion which is passed on to the customer within the price of goods. Conservative estimates suggest that people use between 500 billion to 1 trillion of plastic bags worldwide today and about a million every minute.

ADDRESSING DEMAND FOR PLASTIC BAGS

Plastic do not biodegrade, they photo-degrade into smaller and smaller particles that then contaminate the soil and water, both of which require expensive and difficult clean up processes. This problem suggests that recycling seems like the best solution to the disposal of plastic packaging products in order to keep them from the photo-degrading process. However, less than 1% of plastics get recycled annually and recycling one ton of plastic costs $4,000 which can be resold on the market at $32. The recycling of plastic bags is obviously a highly subsidized (though important) process and points waste policy surrounding plastics to a zero waste solution, or significant reductions in demand.

State governments as well as private companies and non-profit organizations have initiated different plastic policies in order to reduce dependency on and use of plastics. For instance, Ireland's PlasTax program reduced plastic consumption by 90% while generating $9.6 million in its first year enacted. City governments are getting tough on the issue both on US and international soil. Seattle enacted a plastics tax of its own but the strong plastic lobby hindered its impact. In January of 2009, New Delhi enacted a ban on plastic bags, placing a fine or jail term on people caught using plastics. While this seems extreme, the actions of New Delhi show the understood importance of tackling the economically and environmentally unsound impacts of the plastics. Private companies see the benefit of plastic bans as well. Retailers spend millions on plastic annually, and a city-wide plastics tax would decrease their need to expend on these harmful and continuously cost burdensome products. Whole Foods, Walmart, Aldi's and IKEA all have plastic policies in place to discourage generating plastic use and continue to thrive.

In 2010, a survey carried out on the use of plastic in Kenya shown that an estimated 24 million plastic bags are used monthly. These plastic bags take between 20 – 1000 years to bio-degrade and thus have long presence in the environment. Kenya, like most developing countries is yet to have efficient and eco-friendly plastics waste management strategies. As a result, there is widespread plastics pollution across most ecosystems and the situation gets worse each day.

Pollution due to plastic results in major negative environmental, socio-economic and health consequences. From an environmental perspective, plastic bags are to blame for deaths in sea animals through ingestion and suffocation through entanglement. Further, plastic degrade the aesthetic natural beauty of landscapes and lead to visual, noise and thermal pollution. Studies have shown that plastics usually release a very dangerous powdery dust when they break. This powdery dust is known to threat many lives both in water and soil.

From a socio-economic perspective, discarded plastics block sewers and drainages which ultimately cause severe flash floods in major urban and peri-urban areas. This results to destruction of infrastructure and subsequently innumerable financial sufferings thereby adding financial burdens to already constrained national and county governments.

Source:http://www.cloud9service.com/single-post/2016/06/07/A-Parking-Lot-Flood-Can-Be-Caused-By-Clogged-Storm-Drain

Figure 1.12: Plastics block drainages which ultimately cause severe flash floods in major urban and peri-urban areas.

Further, grazing animals ingest plastic bags due to their indiscriminate feeding habits and once ingested, the animals suffer indigestion, immune suppression, ruminal obstruction, loss of weight, depression, reduced milk yield and bloat. This leads to increased veterinary costs to care for the animals most of which die due to serious health complications. In turn, the pastoral communities are negatively affected as they solely depend on livestock for their livelihoods.

Dumped plastics containers filled with rainwater may become breeding grounds for mosquitoes and the resultant effect is increased risks of transmission of various infectious diseases (such as yellow fever, malaria, leishmaniasis, dengue, and human African trypanosomiasis among others). Additionally, plastics create suitable breeding grounds for various vectors, especially those that transmit parasites and pathogens from an infected animal to another. This usually results to a very sick society (both among the humans and animals). This situation is usually aggravated by poor sanitation system whose state is compounded and negatively affected by plastic litters.

Additional, plastic bags in freshwater ecosystems are a good media for the growth of algae which attracts snail vectors which transmit snail borne diseases such as bilharzia and schistosomiasis. Furthermore, open burning of plastic bags besides contributing to air pollution, emits "dioxins" and "furans" as by-products. These by-products are persistent organic pollutants in the environment and are potentially carcinogenic to humans. They are also responsible for acting as "endocrine disruptors" that affect the reproductive system of humans and other living organisms. More so, in dumpsites, plastics trapped in municipal solid waste create foul smells due to their non-biodegradable nature and are a favorable habitat for flies and rodents which are pathways for pathogenic viruses and bacteria.

Therefor, there are countless environmental, socioeconomic and health consequences that have prompted many governmental and non-governmental agencies to respond to the ever increasing plastics challenges. In order to manage the plastics menace; different governments have banned the use, manufacture, and importation of all plastics especially the plastic bags (carrier and flat bags).

In pursuit of the success of the implementation of the bans, governments have developed strategies covering publicity, compliance enforcement and scientific information mobilization. Responsible authorities can take the advantage of the following activities in order to enhance the effectiveness of the bans.

- Talk shows in media houses
- Use of social media platforms
- Consultative meetings at national and regional forums
- Development and dissemination of awareness and publicity material
- Adverts, supplementary and notices in specified print media
- Developing documentaries and animations on plastic bags
- Public and regional awareness/sensitization meetings and identification of alternatives to plastic carrier bags

In the Compliance Enforcement Strategy, the responsible authorities should serve notice to manufactures, importers and large consumers such a supermarkets to submit their compliance plan to the plastic ban and a declaration of stocks held by them by the time the ban takes effect. Other strategies that can be effective in the implementation of plastic bans include liaising with county governments for cooperation in effecting the ban, sensitizing the magistrates on the plastic ban notice, drafting of criminal charges applicable to the offenses relating to non-compliance to the ban and undertaking regular surveillance and enforcement on the plastic ban.

Additionally, under scientific information mobilization, responsible authorities should collect, collate and analyze existing data on plastics, collaborate with UN Environment and other stakeholders for data on effects of plastic bags across sectors, document experiences and lessons learnt from countries, develop a database system for archiving data, undertake further studies on plastics in collaboration with stakeholders including UN Environment, County Governments, abattoirs, manufacturers and finally map entities dealing with plastics manufacture and importations.

POLICY INSTRUMENTS FOR PLASTIC MANAGEMENT

Management of plastics is a general problem globally. This is as a result of inadequate framework in solid waste management. According to UNEP (2005) viewpoints, a good waste management framework should provide guidelines in standard improvements, efficiency and collection of plastic in the environment.

Policies on plastics are found on all six populated continents. These policies use instruments (tools) that range from regulatory, to environmental taxation, to non-regulatory (voluntary). There two ways in which plastic policies can be explained. The first way involves typology of three distinct instruments that influence collective action concerning environmental problems. The second ways involves breaking these three instruments into specific plastic policy instruments.

Typology of Plastic Policy Instruments

When selecting and designing the kind of plastic policy to implement, it is advisable to first make a decision which policy instrument to use. Most often, there are three major types of policy instruments that are used in influencing and creating collective actions in plastic policy. These include the regulation, market-based, and the non-regulatory instruments as illustrated in *table 1.1* below.

Table 1.1: Illustrating key policy instruments that are used to regulate plastics consumption

Regulation	Market-Based	Non-Regulatory
• *Command-and-control* • *Standards*	• *Environmentallev-ies/charges* • *Deposit-refund systems* • *Subsidies* • *Permits*	• *Voluntary Agreements* • *Informational Campaigns* • *Environmental Covenants*

Source:http://climatepolicyinfohub.eu/non-market-based-climate-policy-instruments

*Regulation:*this is also known as command-and-control (CACs) regulation. Regulation is "a directive to individual decision-makers requiring them to set one or more output or input quantities at some specified levels or prohibiting them from exceed (or falling short of) some specified levels". This instrument has two proposals for the policy guidelines with their advantages and disadvantages. The policy guidelines recommend for the outright bans for plastic bag an enactment of minimum film thickness standards. According to UNEP (2005) report, regulation sets the standards on "what, how, when, where, and how much to produce, consume, emit, and clean up". The large amount of control is typically targeted at the largest risks and problems.

These rules directly control activities of individuals or actors, rather than the outcomes of those activities. The rules are mandatory and present limited options for action, and also have legal sanctions or penalties for non-compliance. Based on this rules, bans on the manufacture, distribution and trade of plastic products have been applied in a number of developing countries like Haiti, Ghana, Bangladesh, and Rwanda with a varying degree of success.

Regulation requires a large amount of direct government control. The government (or overseeing institution) must be vigilant in monitoring and enforcement, as well as have the capacity to do so. It is also important to establish punishment for non-compliance, in order to assert authority. Actors are motivated to comply with the regulations, or else face fines or penalties for non-compliance.

One of the main advantages of regulation is that the regulator (i.e. government) has a reasonable degree of knowledge of the expected and predicted outcomes. Regulation also guarantees equality. This means that the same rules apply to everyone, placing everyone on the same playing field.

One of the main disadvantage of regulation is that direct monitoring and enforcement of every actor may be time consuming and costly for the overseeing body. Established rules create little or no incentive for compliers to develop means or set goals beyond those required. Also, businesses may be at a disadvantage compared with overseas competitors who aren't required to comply with the regulation.

Market-Based: This is also known as economic instruments, market-based instruments create economic incentives for parties to comply and actin accordance with law, and go beyond what is required by laws and regulations. In other words, this policy guideline includes levy and subsidies. The recommendation on levy indicates that a well-designed levy has considerable potential to address the problem of wasteful production and consumption of plastics in a country. Market-based instruments are appropriate when the impact of the instrument influences the behavior to be influenced as closely and directly as possible.

Economic instrument can have two purposes: (1)to cover the costs of environmental services and expenditures, or (2) in centivizea change in behavior. According to UNEP (2005) explanation, it can be concluded that economic instruments "aim to correct failures and reinstate necessary condition for sustainable development".

However, a large segment of stakeholders may believe that recycling minimizes the need to reduce production and consumption of plastic products. The policy on subsidies indicates that recycling schemes all over the world rely on government subsidies because of limited market for recycling products. The targeted subsidies for sectors such as those involved in recycling, development of a managed disposal system and production of environmental friendly alternatives have great potential.

Compared to the regulation, this approach usually requires a smaller amount of direct state control. However, some government monitoring is required for maintaining revenue records, enforcement by authorities, and administration of the levy.

When it comes to the advantage of using this instrument, it is worth noting that this approach is much more flexible. A charge establishes a 'willingness to pay' for the good or service- consumers can avoid paying the levy by not consuming the priced good.

On the other hand there are some disadvantages associated with the use of economic instrument. For instance, some researches must be conducted to determine the appropriate charge or fee. It is important to distinguish between a levy (fee) and a charge. A fee is paid for receiving a good or service, while a charge is paid indirectly, like a tax, and may not be clearly displayed. It has been argued that taxes can have distributional consequences, such as placing heavier burdens on the poor and distorting goods prices.

Non-Regulatory: Also known as voluntary, it involves awareness campaign and voluntary agreements as policy guidelines to be adopted in a country. Studies indicate that awareness campaigns and environmental education are very important support instruments for both pre and post-consumer measures. The suggestion of voluntary agreements for code of practice for retailers and manufactures can produce considerable success in tackling the problem of plastic in a country. Voluntary approaches are based on the idea that participants can decide to commit themselves to go beyond; it requires parties to create specific rules applicable to the participants who commit to voluntary reduction or environmental improvements.

A popular choice of non-regulatory instruments is voluntary agreements. There are two classifications of voluntary agreements: individual (targets are set individually), and collective (targets are unilaterally agreed upon). Voluntary does not mean without obligations-there must be realistic targets, mutual trust, communication, and transparent monitoring.

Participants of voluntary awareness also attain costs for involvement. Because there may be no (or little) government intervention in the voluntary awareness, participants are responsible for any and all transaction, start-up, administrative costs, etc. These costs can include: drafting the agreement, data gathering, data checking, distribution, monitoring, revisions, and sanctioning; these costs grow with the number of participants.

Non-regulatory approaches do not contain much (if any) direct control by the state, but the government will enter into a voluntary awareness if it is determined to be more efficient than other policy alternatives (i.e. regulation).

In terms of advantages, it is worth noting that voluntary approaches have benefits for retailers by allowing maximum flexibility in achieving agreed outcomes, avoiding stricter regulation options, cut costs, have self-determined goals, gain a 'responsible' reputation, and establish targets beyond the status quo. Governments are subsequently relieved of regulatory overload.

In terms of disadvantages, it is worth noting that agreeing on voluntary action would require collaboration between retailers, which can be difficult to achieve and may be seen as interfering with market pricing mechanisms. Also, in absence of legal enforcement and regulation, there is the possibility of free-riders and usually few penalties.

Plastics Policy Instruments

Governments around the world have implemented a range of initiatives of policy instruments to modify consumer behaviour and consumption of plastic bags. The three policy instruments categorized above can be broken down into specific plastic policies: a ban, a tax, and voluntary retailer action. While there is consensus that plastics consumption has damaging effects, there are divergent views as to how consumption should be addressed.

Plastic Bag Ban ("Regulatory"): According to UNEP (2005) viewpoints, a regulatory approach to addressing the plastics problem is an outright ban. The UNEP (2005) report states that experience in other countries has shown that bans are the most effective policy instrument in areas with "high vulnerability to litter, single-use character, low (if any) price, and poor recycling feasibility".

A ban can substantially reduce the amount of plastic produced and consumed, littered, and disposed into waste. The ban legislation must contain specific and binding guidelines and rules to be followed. In addition to the required legislation, additional factors must be considered. Some of these factors are "the phase our period before implementation of the ban, punishments for non-compliance, enforcement, monitoring and administrative responsibilities".

A regulation ban on plastics does not allow individuals the option of whether or not to use plastics, therefore taking away the willingness to pay or facilitate consumer choice. There must also be exemptions from the ban for health and safety reasons. These exemptions can include fresh/raw meat and produce products, chemicals, and airport duty-free bags. Issues that may arise from a ban on plastics include an inconvenience to consumers if they forget to take a carrying bag with them, and some consumers may switch to alternatives which also may be unsustainable (i.e. paper bags). Other examples of regulatory plastics policy could be a minimum thickness standard, mandatory recycled content, mandatory participation in a product-or-producer responsibility program, etc. are accompanied by a minimum thickness standard, meaning that the ban only applies to plastic bags below an established thickness; bags above the legal thickness are allowed.

For instance, a country Like Bangladesh implemented a complete ban on plastics in 2002 after littered bags were blamed for contributing to massive floods in 1988 and 1998, which affected ⅔ of the country. Bangladesh banned the distribution and use of plastics after they were discovered to be blocking the drainage systems, which led to increased damage and causalities. The ban eliminated the 9 million bags used every day, and today use of plastics can result in a £9,000 fine in Bangladesh.

Plastic levies (market-based): a market-based approach to addressing plastic issues is usually in the form of environmental levies and taxes.

Tax on goods is an instrument that is meant to gradually change consumer behavior in consumption. It is a market-based approach, which is also known as an economic instrument. This approach creates an economic incentive for individuals to comply with a law and works when the impact of an instrument changes the behavior to be influenced as closely and directly as possible. In order for a tax to be fair and effective in changing behavior and ultimately reducing use of consumable goods, the charge must be at an appropriate level.

The tax can be prescribed to be collected either from the supplier of plastics or the consumers, but a set charge must be applied for consumers who chose to use plastics. The revenue generated f rom a plastics tax may then be earmarked for helping to solve problems associated with plastics use, or help fund other environmental issues. According to UNEP (2005) suggestion, where the generated revenue is allocated can determine the effectiveness of the tax. An appropriate charge would be enough to cover the cost of production, distribution, consumption, and disposal. A tax is most effective if it is implemented at the national level.

One good example is South Africa which enacted a tax in 2003 after initial attempts at a complete ban in 2001 became impossible to implement. Nowadays, the tax and ban are run concurrently and varies with the size and thickness of the plastics. For instance, those plastic bags that are less than 30 microns are banned, while those above 30 microns are taxed at 3Rands per bag. The tax is imposed at the point of purchase (retailers), and failure to comply can result in a fine and/or prison. After closing loopholes within the legislation, the use of new plastics fell by 80-90% within the first three months.

Voluntary Plastic Action ("Non-Regulatory"): Voluntary initiatives could be applied to plastics use; voluntary levies or bans can be structured by a scheme such as a national code of practice for retailers and manufacturers. A voluntary code will commit retailers to reduce the amount of plastics used by implementing reduction-based initiatives, such as;

- Minimum charge for plastics
- Increasing plastics produced from recycled materials
- Reporting total plastics used in a specified period
- Voluntary labeling
- Promoting reusable (i.e. cloth) bags
- Campaigns and incentives to motivate consumers to bring their own bags

Voluntary actions used to address plastics can also be integrated concurrently with other the necessary policies, such as regulations or economic instruments; they can also be put in place more quickly than other alternatives, allowing for quicker action towards plastics.

The threat of regulation is an important determinant in voluntary awareness participation and effectiveness. Signatories will determine if the threat of plastics regulation is more costly and problematic, then (most) will agree to sign the agreement in order to avoid the regulation. For example, if the objectives of the voluntary awareness are not met (i.e. reduction of 50% within 2 years), then a ban will automatically be imposed. This 'threat' of regulation can be a motivator for achieving voluntary awareness targets.

A good example where Voluntary Plastic Action has been applied is Australia. The country adopted the Voluntary Code of Practice in 2002 (referred to as Code of Practice here within), which aimed at reducing plastic consumption by 50% by 2005. The Code has a supermarket participation rate of 90% and all participants were given a 'kit' with information and tools on how to reduce consumption. As a result of wide participation and commitment, Australia reduced its plastics consumption by 34% since 2002 meeting just over half of its initial goal of 50% reduction.

Top-down Approach: According to UNEP (2010), the top-down approach assumes policy is made at the top and then is given to those at the bottom to follow. Usually, these policy implementations are from the government or managerial positions. In terms of environmental policy implementation instruments, this model favors a command and control or regulation policy. However, to be successful, this policy cannot present such control that it is crippling and needs to be trustworthy and a credible theory of cause and effect to be implemented. The top down approach has been criticized for its failure to take into account the variety and importance of other actors involved in policy implementation, such as businesses and consumers. Keeping these stakeholders involved in the process, the policy will be accepted once implemented.

Bottom-up Approach: Powerful groups such as non-governmental organizations (NGOs), industry, labor unions, and informed individual citizens form the 'bottom' or 'street- level.' This is particularly true during the process of defining environmental problems and during implementing policies. These bureaucrats are on the 'ground' and can be recognized when a policy needs to change or be implemented. These activists can rally and create grassroots movements for behavioral change and then when there is enough support, push the movement towards the government for policy reform.

GOVERNMENT INVOLVEMENT IN REGULATING PLASTICS USE

It is therefore important for the government to be involved especially when it comes to matter of creating and implementation of these policies that seek to address the concerns, to correct the failures in the markets, and to provide the necessary public goods. Government involvement is often required for issues that regard public concern and safety, significant economic, social, and environmental effects, and international issues; intervention is warranted with evidence of concern regarding these issues. For the countries that do not have adequate (or lack) policies or systems concerning plastic consumption suffer from governmental failure to address this issue.

Governments usually have more power, resources, information, and means to address the issues associated with plastic usage than compared to state or local organizations. Government intervention targeted at reducing plastic usage will be more efficient in creating and allocating the necessary resources compared to individual free market would as long as the benefits of government intervention outweigh any negative effects and/or costs. Governments are able to work with a variety of stakeholders, can establish regulatory framework to work within, and can provide funding to assist in making necessary changes.

While the number of countries with a government-implemented national plastic bag policy has steadily increased since the 1990s, there are still many countries (developed and undeveloped) that have no such policies. Although all six populated continents have countries that have taken some actions against plastic bag, legislative or otherwise, a large proportion of countries with national plastic bag policies are concentrated in Asia, Africa, and Europe, whilst North and South America lag behind the movement.

Table 1.2: Illustrating top 20 GDP countries (with and without plastic national policies)

With a plastic national policy	Without a plastic national policy
• China • India • Germany • Brazil • France • Italy • Spain • Taiwan	• United states • Poland • Japan • Australia • Russia • Turkey • Mexico • Iran • United kingdom • South Korea • Canada • Indonesia

Source:https://en.wikipedia.org/wiki/Phase-out_of_lightweight_plastic_bags

One might expect that more developed countries would be leading the way in environmental action, considering they have the resources, information, and means to address environmental issues than compared to lesser developed countries. However, concerning plastic that is not the case. Out of the top 20 GDP producing countries, twelve do not have a national plastic bag policy (see *table 1.2 above*). The policy responses from the top GDP countries have not only lagged behind that of developing countries, but have also been concentrated at the state or local level.

2 PLASTIC PACKAGING IN FOOD INDUSTRY

INTRODUCTION

Food packaging like any other packaging is an external means of preservation of food during storage, transportation and distribution and has to be provided at the manufacturing/production centre. Hence it forms an integral part of the product manufacture/production and has an important function in the distribution of foodstuffs. In today's consumer oriented economy a package is an extremely vital link between the manufacturer of the product and the ultimate user. There is great awareness among the consumers today regarding their right to obtain proper quality and correct quantity of the product at a fair price in an aesthetic and hygienic package. Most countries around the world are predominantly dependent on their agricultural economy, often leading to severe shortages. It is realised that one of the important causes of food shortage is high post-harvest losses and spoilage, estimated to be around 20%. Packaging of foodstuffs becomes the most important element in different countries such as Africa and Asia where the climatic conditions vary considerably (temperature from sub zero to over 50Degree Celsius and relative humidity from 10% to 90%) and packaging and handling facilities are grossly inadequate.

Source:http://www.ebeez.co.uk/kitchen-dining/Super-locked-10.html

Figure 2.1: Illustrating sample of plastic products that are commonly used in food packaging.

Packaging as an effective tool of marketing has created a new standard of living in the developing countries. For instance, during the past decade in India, there is almost a packaging revolution with the availability of varieties of packaging materials, thus generating better packaging consciousness in other producer/manufacturing food industries. But unfortunately such realisation is not forthcoming in different food sectors and packaging techniques. There are three types of packaging.

Primary Packaging (Sales Packaging) is the packaging that contains the product. This category also includes packaging provided to the final consumer by a retailer, restaurant or other service provider to ease the transfer of the product. Examples: plastic cereal bags, beverage bottles, disposable cups and plates.

Secondary Packaging (Grouped Packaging) is the outer packaging used to pack individual products, protect the product, add visual appeal or aid in the stocking of shelves. This type of packaging can be removed without impacting the product's performance. Example: six-ring holder for beverage cans toothpaste boxes.

Tertiary Packaging (Transportation Packaging) is the packaging used to safeguard goods and facilitate product handling during transportation. Example: corrugated boxes, wooden pallets, shrink-wrap.

The most important aspect of packaging is to project a high perceived value that people cannot resist but want to buy the product. Food packaging has evolved tremendously over the years with quality materials, creative designs, interesting forms and functions with the use of advance technology. In order to compete in the global market, food sector needs to position itself effectively to capture the young generation market segment.

Today, there are changing needs and lifestyle in customers, for example prepared food, bottles drinks, preservation of foodstuff for distribution and promotion of sales and marketing. Therefore, there is a need to protect consumer health and promote saleability and marketability of products. Moreover, national capability in packaging technology and packaging trade information should be improved. There is a need for an independent national focal point on packaging technology and design elements of food products.

PLASTIC PRODUCTS IN FOOD PACKAGING

Plastic is one the fastest growing material commodities. Approximately 100 million tons of plastics are produced around the globe annually, which equates to 40 pounds for each individual in the world. The largest use for plastics is packaging, accounting for 34% of the market, followed by construction materials (26%) and automotive parts (18%). Half of the plastic used in packaging is fashioned into containers (e.g. bottles, jugs and tubs), another third is used to form sheeting and films, and the remainder is used for closures (e.g. caps and lids), coatings and other functions.

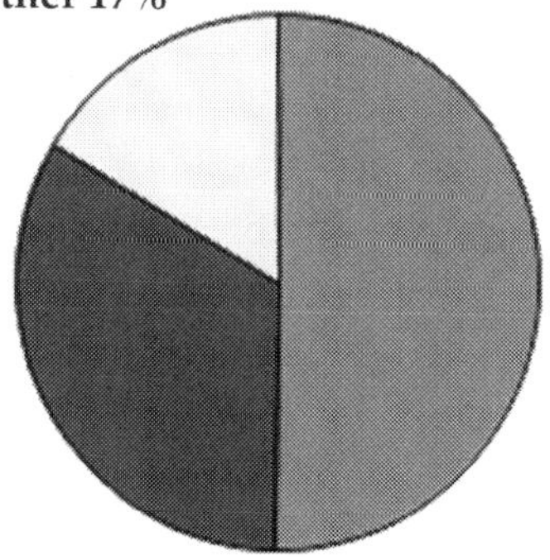

Source:https://www.ihs.com/products/biodegradable-polymers-chemical-economics-handbook.html

Figure 2.2: The Use of Plastics in Packaging.

The largest end-use market for packaging is food preservation. Plastics' durability, versatility and low cost make them ideal for food applications. An estimated 80-percent of plastic packaging is filled by food producers for sale to restaurants, institutions and food retailers. Plastics can be divided into two types: thermosets and thermoplastics. Thermosets are plastics which cure when heated and cannot be melted and reformed. These plastics are difficult and costly to process and as a result, are rarely used in packaging. On the other hand, thermoplastics are widely used for packaging, because these materials can be easily and cheaply converted into an endless array of shapes. Thermoplastics commonly used in packaging include, polyethylene (LDPE and HDPE), polypropylene (PP), polystyrene (PS), polyethylene terephthalate (PET), and polyvinyl chloride (PVC).

Uses and Benefits of Plastic Packaging

Plastic materials in food packaging have many uses and benefits. For instance, it protects the contents from damage due to improper handling and transport problem, and helps to keep the contents safe and to ensure that the product inside reaches the consumer in good condition. Plastic packaging may help retailers advertise their goods, keep sales records satisfactory and prevent the goods from being tampered. It also provides consumer convenience for easily handled, staked, stored and transported. It may reduce waste by dividing food and beverages into individualized portions which reduces leftovers. Instead, plastic package protects, contains and promotes product to communicate a great deal of information.

Plastic packaging must identify and reflect the quality of product. The package should reflect its image at once-through its shape, colour, typeface, illustration or any others unique elements. Besides protection, plastic package must hold the product for safety and cost effective. Indeed, materials and design proven to be safe, including comply will all legal equipment and making optimum use of appropriate technologies. Moreover, plastic packaging should be designed to preserve the product for its intended life span. For instance, appropriate plastic materials, good scaling performance, and ability to operate in accordance and provide abuse resistance to ensure safe distribution.

Plastic packaging must provide information to the ultimate customer and also during transit. On the other hand, it provides make up instructions, handling and use instructions. Ingredients list or part numbers must be included especially for food products. Besides, the product can be displayed visually or by illustration on the plastic packaging besides combined with brilliant marketing and advertising to attract potential customers.

Plastic packaging acts as the "silent salesman" and a way to boost sales. It helps in displaying the brand, promoting information, and company names and address. Products must be well packaged to allow customers to have confidence on it. Besides that, packaging must be eco-friendly and do not harm the environment. It must be considered the after-use possibilities, use materials which have the least possible detrimental effect on the environment, and easily disposable or recyclable.

Popularity of Plastic in Food Packaging

The rise in the popularity of plastic food packaging can be contributed to a number of factors. The following table illustrates some of the key factors that have led to popularity of plastic products in food packaging industry.

Table 2.1: Illustrating key factors that has led to the popularity of plastic in food packaging industry

1. Changes in family structures and dynamics • There has been an increase in the number of single-parent households • There has been a reduction in the size of families (fewer children) • There has been a growing number of women in the workplace • There has been an increase in the number of senior citizens and a rise in the number of single-person households As a result of these changes, fewer people have time to spend on food preparation and more individuals are cooking just for themselves. This has led to demand for food products which can be prepared easily, in limited time, and with modest culinary skills. The versatility and durability of plastic make it ideal to meet consumers' evolving needs
2. Consumers' concerns over product safety and tampering have led manufacturers to increase the number of layers and protective seals on their goods. Plastics, which are often used in film form, suit this demand
3. Technological advancement has significantly changed packaging over the last twenty years • The advent of the microwave oven led to a dramatic increase in demand for plastic films and containers • Advancements in polymer science and plastics engineering have expanded the use of this commodity • Package engineers have combined traditional packaging materials, such as metal and paper, with plastics to create convenient, lightweight packaging materials (e.g. aseptic juice boxes, made of combination of metal, foil and paper) • Composite plastics, which combine multiple layers of different resins, have allowed plastics to be used in new applications. For example, squeezable bottles for condiments are generally composite plastics made of various layers of commodity plastics, engineered resins (e.g. oxygen barriers) and adhesives

4. Plastics are favored by many packaging designers for their versatility, durability and low cost
- Plastics can be processed in a number of different ways to form sheets and films, and foam, semi-rigid and rigid containers
- Pigments and dyes can be added to resins to yield opaque coloured material or transparent coloured film
- Additives, such as plasticizers and heat stabilizers can be added to improve the characteristics of plastics and allow designers to create the ideal package

Source: http://www.brigit-project.eu/detalle_noticia.php?no_id=5480

Attributes of Plastic in Food Packaging

Plastic has a number of advantages over other packaging materials:

- *Low density.* Plastic has a lower density than metal or glass. The result is lower costs to producers for transport and increased convenience for consumers.
- *Shatter resistant.* Plastic is safer during shipping because it will not shatter like glass, thus reducing product waste and decreasing safety hazards.
- *Extremely Versatile.* Plastic can be made into film or shaped into containers. Additives can be incorporated to improve the characteristics of resins. Plastics can be combined with other packaging materials to provide a new packaging system with superior performance characteristics.
- *Reusable and recyclable.* Plastic is durable and can be reused. It is also feasible to recycle many plastic packages
- *Environmental durability.* Plastics are resistant to most environmental conditions (except UV radiation), unlike paper, which is readily degraded when exposed to such conditions.

Plastics have a number of characteristics which make them environmentally unattractive. Plastic resin is a product of fossil fuels. Manufacturing short-lived products, such as packaging, from natural gas and oil is viewed by many as an inefficient use of valuable, non-renewable resources.

Natural gas and oil exploration and excavation disrupts wildlife habitats (e.g. roads, noise, and influx of workers), mars the landscape, and results in subsidence in the extraction region. The consequences of off-shore drilling accidents, such as leaks and blowouts, can be devastating, eradicating fish and bird populations and contaminating miles of beach. The transportation of fossil fuels from excavation sites to refineries can also lead to significant environmental damage, including oil spills, pipeline leaks, and explosions.

Refineries are responsible for breaking down oil and gas into its constituents. These facilities require significant quantities of energy and are major contributors of air and water pollutants. The conversion of petrochemicals into resins and plastic products produces toxic emissions and effluents, and generates large quantities of waste materials. For example, the production of one ton of low-density polyethylene (LDPE), which is considered a relatively environmentally benign plastic in comparison to poly-vinyl chloride or polystyrene, results in sixty-two to ninety-two pounds of organic pollutants. Many of the additives used in plastics to modify or enhance mechanical, physical or chemical characteristics, such as pigments, inks, dyes, plasticizers and heat stabilizers, contain toxic substances, including arsenic, lead, cadmium and endocrine disruptors (e.g. phthalates and nonylphenol).

Technically, most plastic packaging materials can be recycled. However, the infrastructure for physically recycling plastics and the markets for secondary materials are underdeveloped. Secondary plastics have limited end-uses. Recycled materials cannot be used for food contact applications; recycled plastics are typically used to produce construction supplies and vehicle parts. Each time materials are recycled, their strength and weight is decreased and they cannot be used in products which require as rigorous use as the original. The result is a spiraling down effect as each successive recycling further reduces the quality of the plastic and diminishes the possible end-uses.

ELEMENTS AND PRINCIPLES OF PLASTIC PACKAGING DESIGN

Identifying the important plastic packaging attributes in the packaged food market, particularly from the consumer's perspective is important. Today, people are much concerned with healthy lifestyle and especially women who are health conscious and wish to keep themselves fit. There are certain elements and principles of plastic packaging design that should be observed when packaging foodstuffs. For instance, excellent and good packaging design requires thinking on design, through from what it look like, how it hold up in shipping, how it will be displayed on the shelf so that the consumer will interact with it, including post-consumer (recyclable and reusability).

Creative design concepts on food products can attract target audience interested to buy it. Interesting design concepts and the packaging structure, illustration, colours, plastic materials used on the packaging and the product identity plays an important role in food marketing. The first impression especially on these graphic elements can attract people to buy it due to affective and demand factors. People may buy a product due to its strong corporate identity and innovative design on the packaging. Choosing the most appropriate marketing strategy, promotions and positioning can help to increase people's awareness on the product.

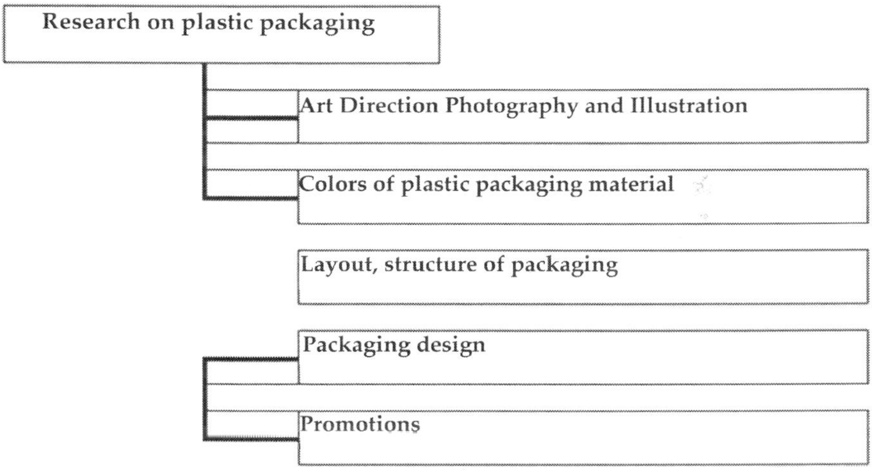

Source:https://www.hindawi.com/journals/ijps/2012/302029/

Figure 2.3: Illustrating important principles in plastic food packaging and marketing.

The role of good plastic packaging is to grab the consumers' attention and create a visual impact. The packaging should capture the customers' interest to a product good with outstanding design and quality materials, enough information provided, unique and create confidence to the customer. Instead, the product needs to dress good and look their best to get noticed in today busy marketplace.

Interesting plastic packaging design and unique forms can add value to the product. Excellent packaging designs are kept for sentimental reasons and are used as collectables for display. Hence, developing new plastic packaging can be expensive because it involves graphic and structural design, production, customer testing, marketing strategy and others.

Furthermore, plastic packaging decisions should take into account environmental considerations. A package that creates problems to the environment can have negative effects and could cause governmental concern of environmental bodies, NGOs and others. A good plastic packaging must avoid over packaging and uses recyclable materials as much as possible.

Additionally, extended producer responsibility (EPR) is another environmental principle where plastic packaging material producers focus on the various impacts plastic products are likely to have in the environment. The concept of EPR encompasses the entire lifespan of plastic products used in food packaging. Such strategies encourage producers to think beyond simply marketing and selling their goods, compelling them to consider the effect their products may have on the environment after the consumer has discarded them. Under the current systems, provincial and federal governments are responsible for setting and enforcing general environmental standards and policies for environmental protection and conservation. The provinces set legislation and regulations related to the use of provincial land and natural resources. The provinces' jurisdiction also includes the governance of municipalities, who are responsibility for solid waste disposal. The federal government has jurisdiction over inter-provincial and international environmental matters, including emissions and effluent expelled as a result of manufacturing processes.

The public plays a role in both plastic waste production and waste management. As consumers, the public is responsible for waste generation and as taxpayers, the public is charged taxes to support refuse management systems.

Under EPR initiatives, the role of municipalities in plastic waste management is typically reduced. The nature of their participation is dependent on the policies adopted. Under some EPR configurations, municipalities are contracted by producers to collect and treat plastic waste materials, while under other systems the producer, retailers or contracted private waste collection companies take responsibility for the physical management.

The role of senior governments remains that of policy developers and enforcers. For instance, in U.S.A., either the provincial or federal governments would be responsible for establishing extended producer responsibility policies and ensuring the actors in the product chain fulfill their assigned obligations.

Successful EPR programs require all the members in the product chain, including resource extractors, raw material suppliers, packaging producers, fillers, distributors and retailers, consumers, government (municipal, provincial, federal), to take responsibility for their behaviour. However, due to the complexity and range of actors, it is necessary when creating EPR policies and programs to delegate one of the participants with the explicit responsibility for managing the product at the end of its useful life.

When producers assume responsibility for the plastic packaging, the costs they incur are incorporated in the final price of their goods. In addition, because of their central role in the product chain, producers have considerable influence over the other actors in the chain and can use their position to encourage others to accept responsibility for their actions. By including the environmental costs of production and consumption in the prices of products, consumers become aware of the costs of plastic waste management and are able to make informed decisions regarding the goods and services they purchase. Consumers also play a role in collecting, separating and returning plastic waste materials.

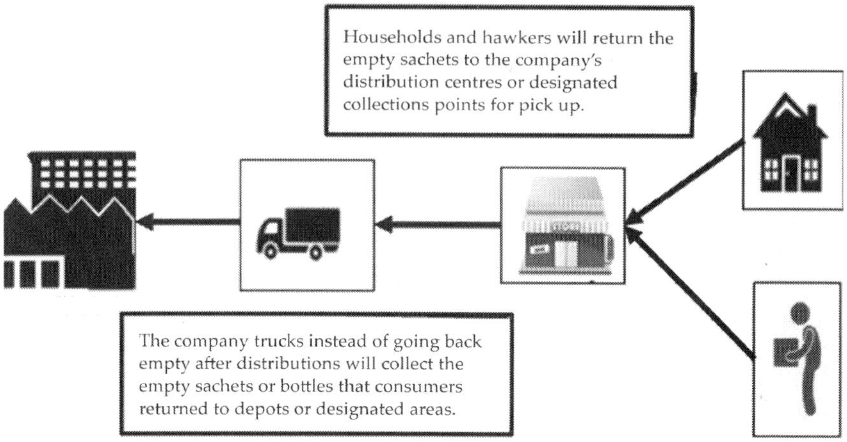

Source:http://www.mdpi.com/1660-4601/12/8/9907

Figure 2.4: Framework for Plastic Waste Management through Extended Producer Responsibility.

According to a statement given by UNEP (2006) "A concept by which industry, governments, and consumers assume a greater responsibility for ensuring that the manufacture, use, reuse, recycle, and disposal off plastic packaging material has a minimum impact on the environment. This includes prime responsibility by industry to design packaging according to the 3Rs principles, take steps to divert packaging from disposal, actively use recovered materials, and ensure packaging is properly handled if it must be disposed. Governments have a responsibility to promote plastic packaging stewardship and to encourage the widespread recognition and adoption of the principles as outlined. Consumers on the other hand have a responsibility to make appropriate packaging choices when purchasing products and, where facilities exist, to divert packaging from disposal".

PLASTICS PACKAGING AND ENVIRONMENT

Most plastic products used in food packaging are usually synthetic substances consisting very large molecules that are produced through chemical reaction. Plastics are one of the organic polymers that have high molecular weight. They are usually synthetics and almost all of them are made from petroleum. Packaging plastics can be remoulded, extruded or processed into others forms including solid object, film and filaments. The structural molecules of the polymers that are plastics are linked together by chemical bonds. Plastics can be divided into two which are thermosetting or thermoplastics materials. The difference between the thermoplastic and thermosetting is their characteristics. The thermoplastic is flexible and becomes soft when heated and hard when cooled while thermosetting is hard and brittle and cannot be heated and remoulded like thermoplastic when heated.

Plastics are widely used because of their properties that are easy and low cost of manufacturing and processing. However, over 60% of solid wastes produced worldwide come from households and a large amount is attributed from single-use packaging plastic products. Most of these plastic packaging materials usually have a very short cycle (less than one year) and this gives continuously bad effects to the environment. To overcome this problem, the 3R are alternative ways which are reduced, reuse and recycle. The word reduced means minimizing the total amount of plastic materials that enter into the solid waste systems. This can be achieved through redesigning the production and consumption patterns of plastic packaging materials. The word reuse means transformation of the plastic packaging materials that are no longer needed and making them reusable again. But reuse also has limitations. This is because most of plastics materials that are used in food packaging are not designed to be reuse again due to their impurities and contamination. Recycling of plastics after used is possible but the plastics product for example plastic bags are rarely to be recycled. The technology of recycling plastics such as plastic material collection and sorting is usually time-consuming and very expensive.

STRATEGIES TO COPE WITH PLASTIC CHALLENGES IN FOODPACKAGING

Source Reduction

This is one of the most commonly accepted strategies towards reducing various effects of plastic in the environment. It involves reducing the amount (volume and weight) or the toxicity of materials entering the waste stream. This strategy is a proactive means of dealing with plastic packaging waste, rather than reactive like recycling, incineration and landfilling methods. Producers engaged in source reduction activities are not only creating less waste material, they are also reducing the amount of pollution produced and energy consumed as a result of the extraction and processing of raw materials.

Common methods employed by packaging manufacturers to reduce the amount and toxicity of plastic packaging materials include:

- Eliminating unnecessary plastic packaging,
- Converting to bulk dispensing systems,
- Light weighting (reducing the volume and/or weight),
- Use of non-toxic additives,
- Reformulating or redesigning products to eliminate or reduce plastics packaging materials required,
- Use of more durable and reusable plastics,
- Packaging products in larger containers, and
- Use of concentrates,

Unfortunately, despite source reductions placement at the pinnacle of the waste management hierarchy, it is the least used refuse minimization technique. Reduction is difficult to measure and manufacturers rarely receive recognition for their efforts from governments or consumers. Most plastic manufacturers around the world usually have little financial incentive to engage in source reduction. In fact, there are a number of factors which make source reduction unattractive to producers, some of which may include;

- Offering funds for research studies and development to reevaluate ways to alter plastic product or packaging design,
- Increasing and providing consumer education programs to promote the benefits of their revised product or packaging,
- Creating and enlightening plastic products suppliers,
- Assuming the risk that consumers may not respond to their new product or packaging.

While producers involved in source reduction incur additional costs and bear financial risk; manufacturers using recyclable packaging receive credit from both governments and consumers for being environmentally friendly, while municipalities and taxpayers foot the bill for recycling.

Over the years, recycling has gathered significant political and public support. This can work against source reduction, as some of the most lightweight packaging materials on the market are composed of multiple materials or composites plastics, and cannot be recycled. The question becomes which is environmental better – lightweight, compact packages which can only be incinerated or landfilled *or* heavier, more bulky packages which can be recycled?

It is important to note that reduction is not always the best ecological option. Source reduction is not necessarily environmentally superior if light weighting consumes more energy or compromises the performance of the package. More raw materials and energy go into manufacturing products, than packaging. If the product is damaged or spoils because it was packaged improperly, more natural resources are wasted in comparison to the small amount of packaging material conserved.

Reusing

Refilling or reusing existing plastic bottles, containers and tubs is an effective way to minimize the consumption of virgin materials, conserve energy, reduce the expulsion of pollutants, and decrease the waste entering landfills and lower manufacturing costs.

The three common packaging reuse systems are as follows;

- Consumers return used packaging to the retailers, recycling depots or reverse vending machines. The packaging is then collected by the producer who washes and refills the bottles. Take-back programs and deposit-refund programs can be used to encourage participation in reuse.
- Packaging is sold to an intermediary that washes containers and sells them to a producer to refill.
- Producers sell concentrated versions of their product in smaller, lightweight packaging. Consumers dilute the concentrate with water in a reusable bottle.

Reusable systems are only beneficial if they use less energy than their non-refillable counterparts. The amount of energy reusable containers consume is dependent on the proximity of the refilling plants to collection facilities and the number of times the packaging is reused.

Because these packages must withstand multiple uses, fillings and trips between the producer and consumer, reusable containers tend to be thicker and heavier than non-recyclables. As a result they require more resin and energy to produce, and consume more fuel during transport. Therefore less energy is consumed when the distance between the retailer and plant is minimized and the container is reused the maximum number of times feasible.

There are several disadvantages to reusable packaging systems. First, reuse systems are not appropriate for all products. Packaging contamination and hygiene concerns often limit the ability to reuse many food packages. Second, the initial cost of converting to reusable containers is significant. Using reusable bottles requires producers to invest in washing, specialized filling and waste water treatment equipment. However, in the long term, there is the potential for substantial savings (reduced energy and raw material consumption). Third, many retailers are not interested in selling refillable containers. Under most refillable programs, grocery and convenience stores become the point of return for bottles and containers. This means retailers must have enough storage space to house returns, pay for additional staff hours to sort and stack refillables, and take additional precautionary measures to ensure conditions remain sanitary. Many retailers are unwilling to bear the costs and hassles of refillable programs.

Recycling

There are numerous ways in which plastics materials used in food packaging can actually be recycled.

These include;

Primary Recycling (Pre-consumer Recycling) involves converting homogenous, uncontaminated plastic waste material into resin or pellets. Primary recycling conserves the most energy of all recycling processes and yields a product with comparable properties to that of the original. This form of recycling is predominantly used by industry to recycle plastic scraps remaining after manufacturing.

Secondary Recycling (Mechanical Recycling) involves converting plastic waste material, typically post-consumer, into new plastic products. For sanitary reasons, food products cannot be composed of secondary materials. Most recycled plastic converted via this method is used in construction materials and automobiles.

Tertiary Recycling (Chemical Recycling) involves chemically breaking down plastics to produce fuel which can be used to generate energy or manufacture chemicals. Processes such as pyrolysis, glycolysis, and hydrolysis are techniques used to chemically recycle plastics. Tertiary methods are very expensive and not widely used.

Quaternary Recycling (Energy Recovery) involves recovering the energy stored in waste plastics through incineration. Highly contaminated, low quality and non-recyclable plastic waste are best suited for waste-to-energy processes.

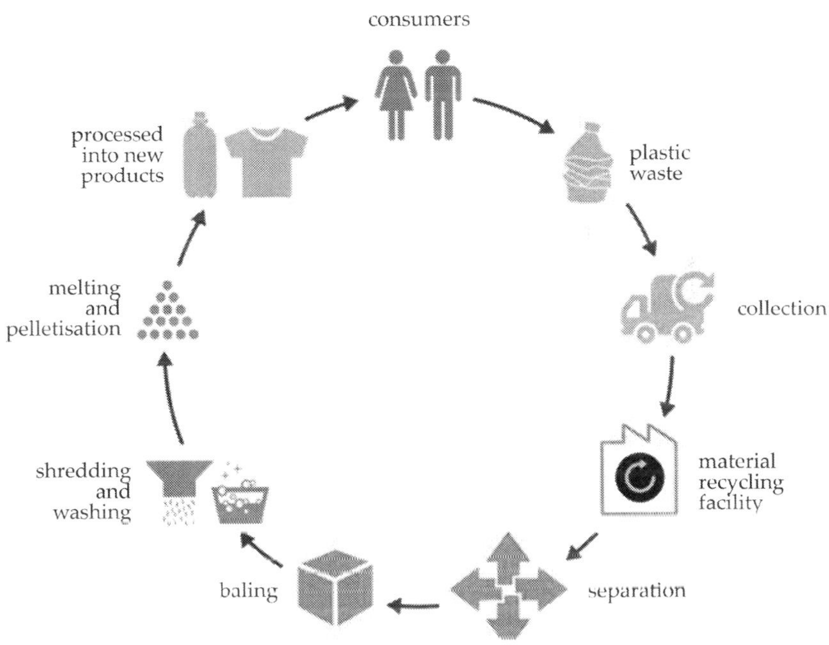

Source:http://www.sita.com.au/commercial-solutions/resource-recovery-recycling/plastic-polystyrene/

Figure 2.5: A typical process of recycling plastic waste materials.

The process of converting fossil fuels into resins represents the greatest energy expenditure in the fabrication process. Using recycled plastics in products, rather than virgin resins, can save between 85 and 90-percent of the energy used in the production of plastics goods. Furthermore, recycling reduces the quantity of packaging materials entering landfills and reduces reliance on virgin inputs.

While there are many benefits to recycling, it is not as effective as source reduction and reuse at addressing environmental problems. Recycling degrades the polymer structure, reducing the mass, strength and integrity of resins. As a result, recycled plastics generally are used only in products which are less demanding than the original. If products made of secondary inputs are themselves recycled, the quality will be degraded even further, until eventually it is unusable. For instance, after three recycling the impact strength of polystyrene is reduced by over 34% and its mass is reduced by 27%.

At this time, it is not economically or technologically feasible to recycle all packaging materials. Packaging of mixed resins (e.g. some squeezable condiment bottles) and mixed materials (e.g. aseptic juice boxes) is difficult to recycle into new products.

Currently, the recycling of plastic packaging is fairly limited. A successful plastic recycling market, like any market, requires both supply and demand. Most municipalities around the world are responsible for operating collection and sorting programs. Collected plastic waste materials typically are sold to recyclers who convert them into a useable form (e.g. pellets, flakes) for use in their own facilities or for sale to plastic fabricators. The extent of municipal recycling initiatives is dependent on availability of funds and the demand by recyclers for various plastic waste products. Plastics are one of the most expensive waste materials to collect and sort because they are bulky, lightweight and have a low resale value.

At this time, plastic processors demand for secondary materials is weak. The price of recycled resin is higher than virgin, the quality of recycled plastics is inconsistent, consumer demand for recycled products is low, and no legislation exist to encourage producers to use recycled materials. For the recycled plastics to be competitive with virgin resins, a dependable supply of sorted waste plastics and investments in recycling technology is required. Unfortunately, most countries are usually reluctant to invest in recycling infrastructure until the supply of recyclables is steady, while municipalities are leery of expanding collection and sorting programs without a reliable demand.

CASE STUDIES ON MANAGEMENT OF PLASTIC USAGE IN FOOD PACKAGING

Statistically, about US$ 400-500 billion is used in plastic packaging industry worldwide. This includes production costs to marketing of the plastic packaging materials. The biggest players in plastic packaging industry are China, Brazil, India and Russia. If including only their market, they cover 30% of the whole market.

There are several factors that contribute to this speedy development and increase in the value of this industry. Some of these factors include;

Increase in consumption by population has increased the demand for plastic packaging items of purchase

Increase in the demand of the consumer goods has led to an increased demand of sophisticated plastic packaging

Exposure of markets to the rural areas has led to an increase in purchase of small sachets

Consumer-base has gained access "Fast Moving Consumer Goods" which has played a significant role in giving a leap to plastic packaging industry

Plastic packaging costs has increased with the increase in middle class families

There are different types of plastic packaging materials. The material to be chosen depends upon the nature of the item and the need of the customer. Mainly materials like plastic bags, plastic containers, plastic sachets, and plastic flat bags among others are used for packaging. Different countries have specific guidelines and systems for managing the use of plastic products in food packaging. The following are case studies that have been identified a cross the world.

India

In India, there are no formal sectors for collecting and recycling plastic wastes. However, there are several informal sectors consisting of plastic waste pickers, buyers and dealers. Among these, plastic waste pickers play the main role. The plastic waste collected by them is then transferred and sold to dealers. This is done on instant payment basis. The rate of recycling in India is 56%, which is considerably a high percentage.

Source:http://leadership.ng/2017/05/18/wealth-from-scrapyard/

Figure 2.6: Illustrating a group of plastic pickers scavenging for plastic waste at the dumpsite.

In India, there is no specific law or legislation containing guidelines for the recycling and disposal of plastic packaging wastes. The only law that comes close to plastic packaging waste collection and recycling is "Plastic Waste Management and Handling Rules which were enacted in the year 2011. These rules are mainly focused on single-use plastic wastes. The drawbacks of the system and legislations in India regarding plastic packaging wastes are that there is no proper recognition given to the waste pickers. There are no specific laws enacted in order to deal with other forms of plastic wastes other than those that are used once. This means that there are no proper enactments of legislation and no strict actions are taken against plastic usage in the country. Also there is exploitation of the plastic waste pickers by the dealers and recycling units.

Nevertheless, there is a case pending in the National Green Tribunal, which deals with plastic packaging waste. This case might bring about changes in the policies and legislations on plastic packaging waste in India. It not only deals with the plastic waste generation but also the pollution and intangible waste like smoke.

Germany

Germany introduced the "Ordinance on the Avoidance of Plastic Packaging Waste (Packaging Ordinance) in the 1991."This ordinance puts an obligation on the manufacturers and distributors to collect the used plastic packaging wastes and recycle them without any consideration and free of cost. This included taking back plastic crates, plastic packaging containers, and plastic drums among other waste materials. The waste collected can be either reused or recycled depending upon the condition and nature of the plastic waste.

After the establishment of the ordinance, there was a non-profit organization (Duales System Deutschland AG) that came into effect. This organization was responsible of collecting, sorting, and recycling various plastic packaging materials that were used by different consumers. The manufacturers who were a part of this organization would attach a Green Dot to their products, which would indicate to the consumers that they do not have to send the waste to the manufacturer and instead the manufacturers would themselves collect the waste.

The criticisms of this system are that it has proven to be cost ineffective. The manufacturers complain that the cost of the consumers has fallen. The privatization of plastic waste management has occurred. There is hasty implementation of the system, which leads to the benefit of the manufacturers as they easily get away with not paying licensing fees.

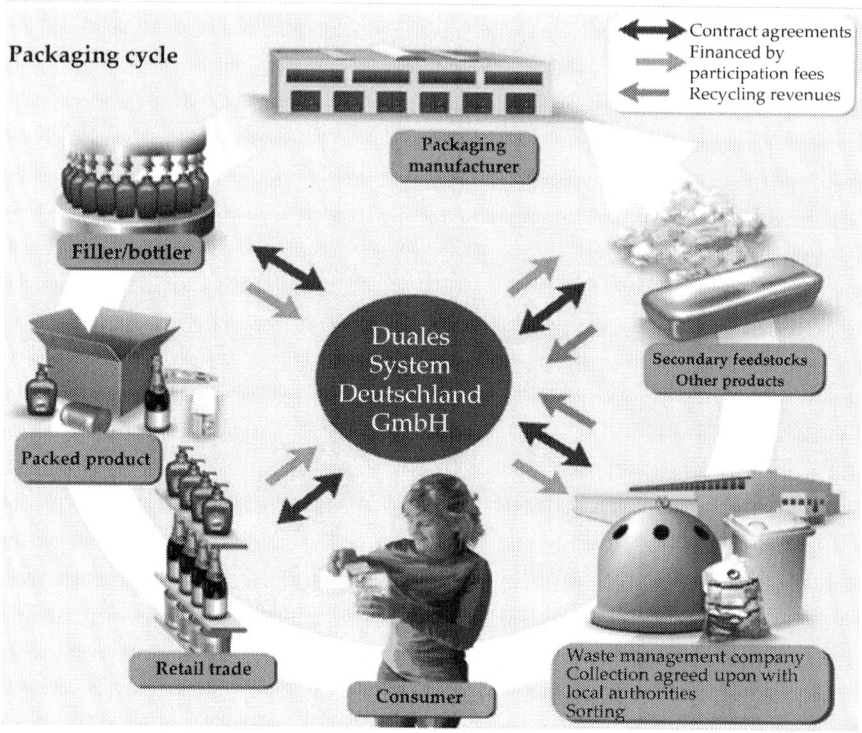

Source:http://nemap-wordpress.blogspot.co.ke/2011/02/solution-to-packaging-waste.html

Figure 2.7: Illustrating how Duales System Deutschland AG works.

Brazil

Brazil, not being very economically well off and facing unemployment does not have a very strong or structured recycling system in plastic packaging waste materials. The percentages of Brazilian Municipalities that have an official waste recycling system are less than 6.4%. Hence there is no formal sector for the collection and recycling of plastic packaging waste materials. Waste pickers similar to India are the main players in collection and recycling of plastic waste materials. The government actively supports them and plastic wastes picking is also recognized as an important economic occupation in the country. The government has come up with a national program referred to as "Carbon

Finance Project", which works at integrating these plastic waste pickers into the local waste management systems.

However, the condition is changing as the top businesses in Brazil have established a Brazilian Business Commitment for recycling which work as a non-profit organization aiming at promoting waste management, especially plastic waste materials.

The drawbacks of the Brazilian system are that it gives preference to landfills and does not take into account alternative methods. Additionally, lack of financial resources hinders the use of incinerators as they are hard to maintain and are expensive to buy.

United Kingdom

Packaging Recovery Note (PRN) system is one of the most important systems involved in regulating the use of plastic materials in packaging industries. According to PRN system, the notes are created when specific quantities of plastic packaging waste materials are recycled and then these notes are bought by brokers who further sell the notes to the manufacturers. These PRNs are not limited and specific to any given location. This system is very cost effective.

European Union Waste Framework Directive is another system that was enacted in the year 2008. This system is used to meet the packaging recycling targets set by the European Union.

The drawbacks and criticism of this model are that the brokers charge fees which reduces the cost effectiveness of the system. It does not target household plastic waste materials. There are little methods to guarantee that the PRNs are used for recycling only. There is a question on the reliability as it complicates the process of meeting plastic packaging wastes recycling targets. This is because no one can point out who is at fault when these targets are not met. This system encourages exportation of plastic waste materials which proves to be a hindrance to the domestic stakeholders.

Japan

The law that governs the recycling of plastic waste materials in packaging industry in Japan is Container and Packaging Recycling Law. According to this law, the consumers of the goods are supposed to sort the plastic waste materials which would be collected by the municipalities. These plastic waste materials are collected and transported by recycling business entities which are selected every year through a bidding process. These recycling entities have to show delivery report to prove that the wastes have been recycled and then only do they get paid for the work. Manufacturers and businesses that use plastic containers have to pay a recycling fee for the cost of recycling.

One of the criticisms against this model is that the collection costs are usually high. The cost to collect plastic containers is 3 times the cost of collecting household solid waste materials. Manufacturers do not get enough incentives to reduce the amount of plastic materials used in packaging various goods and products. Additionally, the recycled materials are sometimes used only as low grade materials which defeat the purpose of recycling.

Hong Kong

Hong Kong has come up with a 10 year plan which works at reducing the rate of landfilled plastic waste materials by 58 percent from the 1998 figures. This plan is known as "Refuse Transfer Station" which works at the promotion of plastic waste materials reduction in addition to reusing and recycling of the waste collected. Here, the government collects the plastic waste from the urban areas and this waste is then stored in containers and in a compact form. This is then transported to the landfill areas where it is dumped. This model reduces the cost of transportation and reduces noise and air pollution that is caused by the transportation.

The drawback of this model is that it focuses on landfills. As Hong Kong is itself a small country, the land used for landfills is limited. With the increase in plastic waste it is unaffordable to buy land for recycling purposes. In other words, it reduces the quality of land.

Australia

The government of Australia recently came up with the Australian Packaging Covenant which has 900 companies under it and targets at recycling 70% of the plastic waste generated over the 5 year plan. The country aims at participation from all the stakeholders of the packaging industry namely the consumers and the manufacturers in achieving sustainable development in the industry.

It plans to provide a sustainable packaging design and to recycle and recover the used plastic packaging materials. Certain states in Australia have come up with landfill taxes and use their revenues to keep up with the cost of recycling. The government also has programs for the promotion and education regarding plastic wastes management.

The National Packaging Covenant, 1999 aimed at cooperation of manufacturers and government where the businesses would help government to achieve recycling targets. If the manufacturers failed then it would lead to strict actions and taxes. Sometimes it leads to the ban of plastic materials in packaging of goods and products.

United States of America

In US the major law that governs the recycling of plastic waste materials is "Resource Conversation and Recovery Act". Other than this, there is no packaging waste law available or any law governing the disposal and recycle of plastic packaging waste.

Every state has their own plastic wastes recycle systems and there is no general system for all. The common method used for disposal is landfill. Other than recycling, plastic waste materials used in packaging sector are collected from bins or garbage cans kept in every locality. There is no system of collecting the waste by visiting every household.

The criticism that America's packaging waste and recycling industry faces is that the recycled plastic materials are usually of low quality and they cannot be further used.

3 PRODUCTION OF PLASTIC PACKAGING PRODUCTS

INTRODUCTION

The plastic industry is one of the fastest growing in the world. In Africa, the industry is growing at a rate of 8 - 10 %. Most of the plastic products produced have short service life, among them are the packaging plastic bags which are very popular because of their versatility, durability and are ease of storage. Most of the packaging plastic products produced in the world are less than 15 microns. This means that they cannot be recycled, thus most of them have become a menace in the environment. Apart from causing visual pollution, most plastic materials have contributed hugely towards blockage of drainage systems, death of marine life and animals through choking. Plastics are non-biodegradable and take 20-1000 years to break down.

Production of plastic products affects various stakeholders in most parts around the world. These stakeholders can be categorised into various groups, which are the manufacturers, the large scale users, environmental and regulatory bodies, the general public, community based organizations and related industries.

Manufacturers: Most plastic manufacturers contribute directly to the problem through over production of thin flimsy plastic bags that cannot be recycled.

Regulatory bodies: Different countries have certain Bureau of Standards that are mandated with the responsibility of regulating the quality of products. Most of the plastic packaging products such as plastic bags produced are usually less than 15 microns meaning that they cannot be recycled. The bread packaging for example is 5 - 7 microns. The Bureau of Standard bodies' role in this case is to ensure standards adherence by the plastic industry.

Environmental organizations: The role of the environmental organizations is to ensure the conservation of environment. They work together with the industries to ensure environmental welfare. For instance, a country like Kenya has several environmental organizations such as United Nations Environmental Programme (UNEP), National Environmental Management Authority (NEMA) and Green Belt Movement (GBM). The UNEP has acknowledged the extent of the plastic materials problem and has spearheaded pilot projects on plastic waste management in different countries worldwide. For instance, the organization launched one project in Kenya on 19th July 2017 and officiated by the Minister for Environment.

Large-scale consumers: The large-scale consumers are those who buy in bulk from the manufacturers for use in packaging. The packaging is either done at the point of sale or at the production stage in the factory. Supermarkets in this case are large consumers as they package all goods bought in plastic materials. Fast food outlets are also large-scale consumers. Most firms who manufacture fast moving consumer goods such as sugar, rice, bread, and potato crisps package the end product in plastic bags.

Community based organizations: Plastic waste management in municipalities has become a duty for every citizen as opposed to being only that of the local authorities. This has become important because the individuals contribute to the litter. Municipal management needs have for long outstripped the capacity of the local authorities to provide sufficient service because often the amounts of plastic waste generated are enormous. Local communities are increasingly taking part in the provision of their crucial municipal service with the objectives of improving the environmental hygiene within their neighborhoods.

Plastic alternatives manufacturers: As plastic packaging materials continue to pose several challenges on the environment, alternatives to plastic products will continue to be considered as viable options. These alternatives include baskets, cloth bags, sisal bags, cardboards, paper bags, biodegradable bags. The *Jua Kali* sector is involved in the weaving of baskets. Women group organizations are also involved in the activity.

The General public: Despite the production of plastic products, the over use and misuse of plastic materials by the public has greatly contributed to the plastic waste menace. Such a problem would not be solved without addressing the public because they are key contributors to the problem.

THE STORY OFPLASTIC PRODUCTION

Plastic production began in 1904, when Belgian chemist, Dr. Leo Baekeland, founder of the Bakelite Corporation, experimented with phenols, a type of thermosetting resin. Baekeland developed a combination of formaldehyde, phenol, and an alkaline catalyst in order to form a thermosetting resin, which, once cured, could not be melted down again with heat to be reformed. Although phenolic resin existed previously, Baekeland's process made its mass-production more viable. Baekeland patented his method of plastic production in 1909.

Concurrently, in an effort to develop new electrical insulation products, the Westinghouse Electrical and Manufacturing Company in Pittsburg, Pennsylvania, tasked the head of the Research and Engineering Department, Daniel J. O'Conor, to experiment with cloths, resins, papers, varnishes, and solvents. By 1909, O'Conor recognized the moisture and heat resistant qualities of the phenolic resin and decided to pursue a working relationship with Baekeland and the Bakelite Corporation. At first, O'Conor was unable to successfully stick the phenolic resin to paper or cloth. He then developed a technique for winding paper around a metal roller while simultaneously coating the paper with warm resin. Once cool, the cylinder could be cut and flattened in order to produce a sheet of plastic-coated paper. O'Conor produced the very first sheet laminate, patented by Westinghouse in 1918, under the trade name Micarta. However, in 1913, before Westinghouse was able to acquire the patent, O'Conor teamed with Herbert Faber, the Westinghouse Technical Sales Manager of insulating products, in order to form their own new company in Cincinnati, Ohio: Formica.

Formica, initially Formica Insulation Company, began producing "V" commutator rings, an insulative motor component, for the Chalmers Motor Company. Two company legends for the origin of the Formica name persist. The official story claims that the material served as a replacement for mica, mica being a mineral which was commonly used as electrical insulation during the early twentieth century. The other story claims that Formica is an abbreviated version of "formally Micarta," Micarta being the Westinghouse-owned trade name. In 1914, Westinghouse attempted to eliminate Formica, its sole laminate competition, by entering into an exclusive agreement with Bakelite, Formica's phenolic resin supplier. With its source for phenolic resin unavailable, Formica sought the aid of a Canadian chemist, Dr. L.V. Redman, to develop an alternate chemical. The new chemical, called Redmanol, combined phenol with ammonia, formaldehyde, hexamethylene, and tetramineammonia, resulting in a suitable replacement for Bakelite's phenolic resin. Between 1919 and 1924, Westinghouse brought three lawsuits against Formica, all of which Formica won. During this period, competition between Westinghouse and Formica largely drove research and development campaigns.

During the 1930s, Formica experimented with decorative patterns. Up until this time the company produced only motor components, tubes, rods, and other molded parts. The first decorative plastic laminates combined different resin mixtures, resulting in a web-like "cracked" pattern, reminiscent of Chinese or Japanese ceramic glazes. During this period, Formica also developed a process by which genuine wood veneers could be laminated with the Redmanol resin, producing a decorative and highly-durable "real wood" surface. In 1931, Formica introduced a "cigarette-proof" laminate which incorporated a layer of aluminum foil, dispersing the heat of a lit cigarette in order to prevent burns. Another Formica innovation during this period was the decorative layer, commonly referred to as the "deco" layer, containing a printed pattern. However, because the Redmanol resin altered the color of the deco layer and darkened when exposed to light, the first patterns were wood grains. The decorative laminate process was not perfected until 1938, when melamine supplanted the Redmanol resin, producing a clear, glossy, and protective coating. In addition to being less expensive than the previous resins, melamine allowed the deco layer pattern to appear unaltered.

Decorative plastic laminates became wildly popular during the post-World War II era. While this rise in popularity can be attributed in part to the post-war housing boom, a variety of other factors prevented plastic laminate from becoming main stream until 1945. Following the innovation of the melamine layer in 1938, a flood ravaged the Formica plant in Cincinnati, damaging the building and machinery. In addition to this setback, partial founder Herbert Faber suffered a heart attack and resigned. Following these blows, the onset of World War II forced Formica to focus on the production of war goods rather than decorative surfaces.

Through the 1930s and 1940s, plastic laminates were used exclusively for vertical wall surfaces, rather than for horizontal applications, such as flooring or countertops. Vertical applications of plastic laminate provided a decorative wall pattern, similar to historic uses of wallpaper or lincrusta, with the added benefit of protective coating and added durability. Plastic laminate was first applied as a horizontal countertop in bars, restaurants, bowling alleys, barbershops, and beauty parlors. The demand for housing following World War II resulted in the first usage of plastic laminate as a countertop surface in the American domestic kitchen and bathroom. Prior to World War II, domestic kitchen countertops where typically wood, ceramic tile, or linoleum. Linoleum was a very popular countertop choice during the 1930s and 1940s for its durability and easy to clean. That it was offered in a variety of decorative patterns also contributed to its popularity. Linoleum paved the way for the introduction of plastic laminate to the American kitchen. Plastic laminate proved easy to install, inexpensive, and highly durable, resulting its explosive growth in popularity. Formica sales went from $4,251,109 in 1940 to over $24,000,000 in 1950. In 1956, American Cyanamid, the supplier of melamine, purchased Formica. The conglomeration earned higher profits by eliminating shipping costs. In the 1970s, Formica became an independent corporation once ore when it was sold by American Cyanamid in a leveraged buyout.

At mid-century, Formica was undoubtedly the leader in plastic laminate production and sales. The product was so pervasive that, like Kleenex, ChapStick, or Xerox, the brand name became the generic name for the product, in this case, plastic laminate. Despite Formica's being the market leader, dozens of plastic laminate manufacturers competed during the 1950s and 1960s. Early rivals to Formica include the Nevamar Corporation, and Panelyte, both subsidiaries of larger producers of plastics goods. Following World War II, General Electric produced its own brand of plastic laminate, known as GE Textolite. Other plastic laminate brands available at mid-century included Ralph Wilson Plastics (later Wilson Art International), Pionite, Arborite, Consoweld, Lamidall, Melamite, Conolite, Reiss, Royalite, and Lamidall. Additionally, smaller companies, such as Daystrom and Beautychrome, produced laminates exclusively for the manufacturing of tables and dinette sets.

FACTS ABOUT PRODUCTION OF PLASTIC PACKAGING PRODUCTS IN EUROPE

According to European Association of Plastics Recycling and Recovery Organizations, more than 1.4 million people in Europe are employed in the plastic industry. Some of these people are involved in production of various raw materials, converting these raw materials in finish plastic products, and in the operational activities for manufacturing of plastic products. By 2014, there were more than 62, 000 companies involved in manufacturing and distribution of the plastic packaging products. The turnover for plastic products in the 2014 was more than £ 350 billion.

In terms of multiplier effect, plastic industry contributes about 2.4 in GDP of Europe. About 4-6% of the total world's oil production is used in the production of plastic materials. Currently, China is the largest producer of plastic materials followed by Europe.

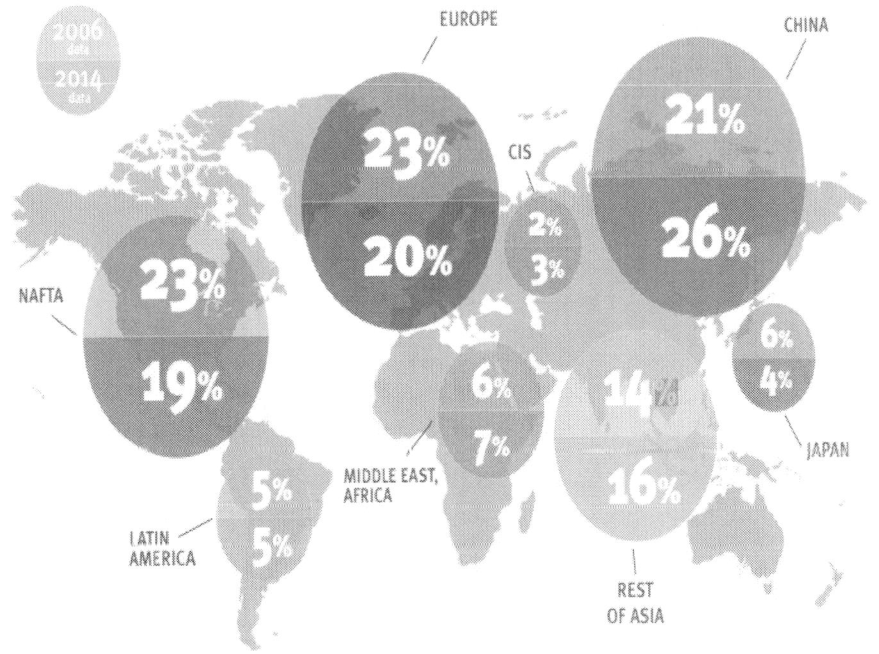

Source:http://www.plasticseurope.org

Figure 3.1: Illustrating 2014 world's production of plastics materials (in metric tonnes).

More than two thirds of the plastics demand in Europe is concentrated in five major countries (that is United Kingdom, Germany, France, Italy, and Spain). The demand for plastic products as of 2014 stood at 47.8 metric tonnes. This includes all plastic materials (both thermoplastics and polyurethanes) and other plastics (such as adhesives, thermosets, coatings, and sealants).

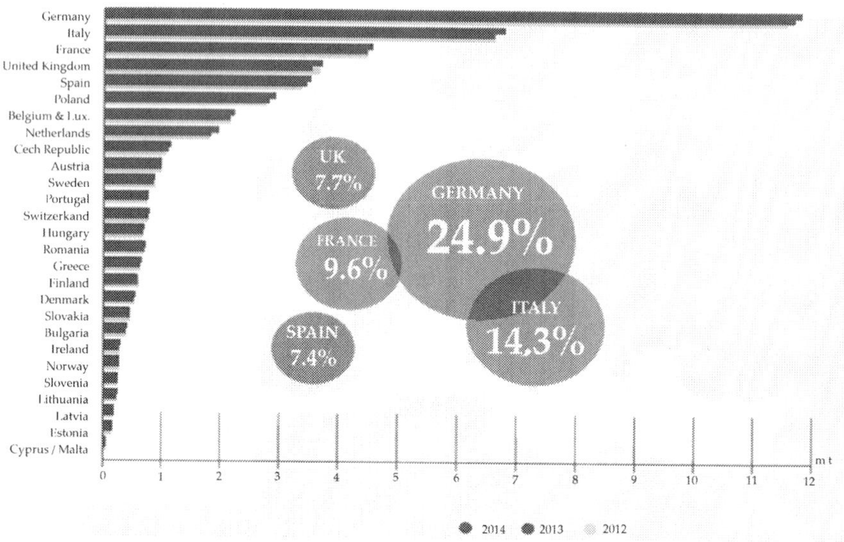

Source:http://www.plasticseurope.org

Figure 3.2: Illustrating the demand of plastic materials in Europe (in metric tonnes per year).

As of 2014, plastic products were consumed largely in the packaging sector (39.5%) and agriculture accounted for the least consumption of plastic products in Europe (3.4%).

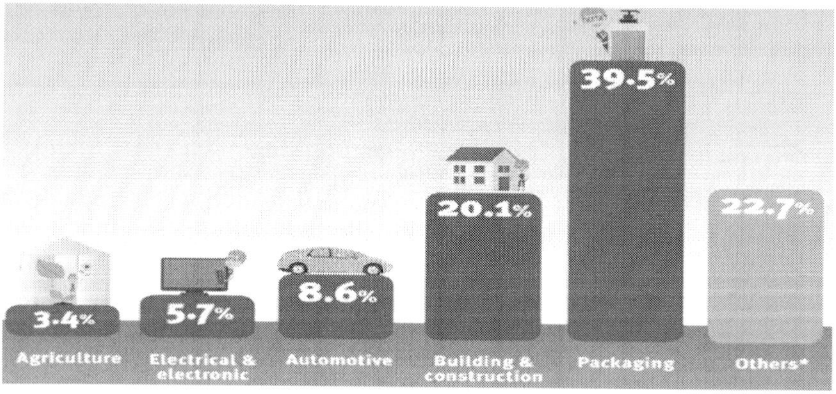

Source:http://www.plasticseurope.org

Figure 3.3: Illustrating some of the major sectors involved in distribution of plastics products in Europe.

In 2014, plastic recycling and energy recovery reached 69.2%. In other words, more that 25 million tonnes of post-consumer plastics waste ended up in the waste upstream. About 69.2% was recovered through recycling and energy recovery processes while 30.8% went to landfill.

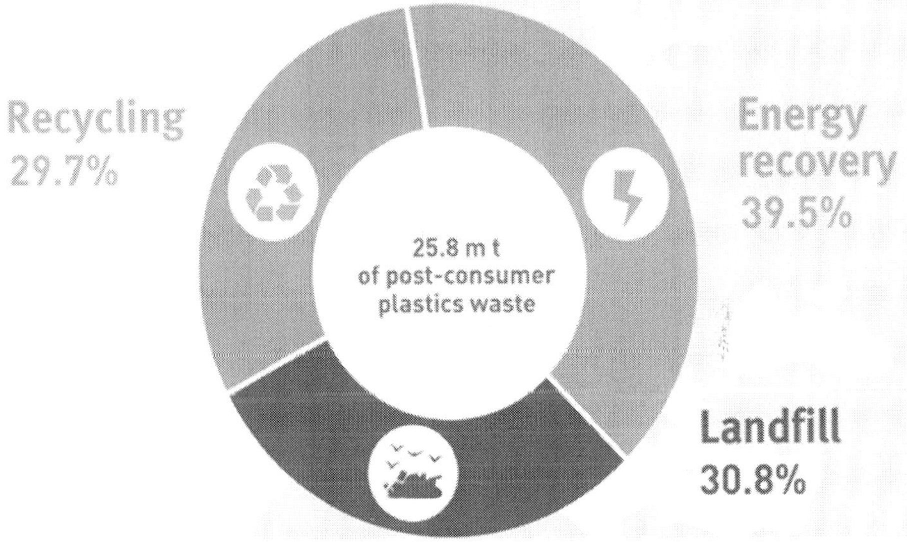

Source:http://www.plasticseurope.org

Figure 3.4: Illustrating treatment for the post-consumer plastics waste in Europe.

Currently, landfilling is still the first option in many countries in Europe. More than 50% of plastics waste is usually landfilled.

FACTORS THAT AFFECT PRODUCTION OF PACKAGING PLASTIC PRODUCTS

There are several factors that influence plastic manufacturer's choice of direction and action and ultimately its organizational structure and internal production processes. These factors which constitute the external environment can be divided into remote environment, industry environment and operating environment. These factors form the basis of the opportunities and threats that plastic manufacturers face in their competitive environment.

Studies have shown that environment is a critical factor for any organisation survival and success. It should be seen as a biosphere in which individual organisation use over the long term as a community project in which to be actively involved. It is a resource to be managed and to be shared, hence the need to effectively manage the value chain system and establish collaborations, partnerships and to get involved in social responsibility to enrich this resource and enhance the corporate image of the organisation.

The remote environment comprises of factors that originate beyond and usually irrespective of manufacturers' operating situation. These factors include; economic, political, technological and ecological factors. The environment presents plastics manufacturers with opportunities and threats but rarely does a single manufacturer exert any meaningful reciprocal influence.

Economic Factors

These factors concern the nature and direction of the economy in which a plastic manufacturer operates. A plastic manufacturer needs to consider the economic trends in the segments that affect its industry. These trends include the general availability of credit, the level of disposable income and the propensity of people to spend, prime interest rates, inflation rates, and trends in the growth of the national product.

Social Factors

The social factors that may affect a plastic manufacturer involve the beliefs, values, attitudes, opinions and lifestyles of manufacturer's external environment as developed from the cultural, ecological, demographic, religious, educational and ethnic conditioning. As social attitudes change, the demand for plastic products also changes. Like other factors in the remote environment the social forces are dynamic with constant change resulting from the efforts of individuals to satisfy their desires and needs by controlling and adapting to the environmental factors.

Political Factors

Political factors define the legal and regulatory parameters within which plastic manufacturers must operate. Political constraints are placed on plastic manufacturers through fair trade decisions, antitrust laws, tax programs, minimum wage legislation, pollution and pricing policies, administrative jaw boning and many other actions aimed at protecting employees, consumers, the general public and the environment. Since such laws and regulations are normally restrictive, they tend to reduce the potential profits of the manufacturers. However, political factors may also benefit the firms they influence. This is through patent laws, government subsidies and product research grants.

Technological Factors

To avoid obsolescence and promote innovation, plastic manufacturers must be aware of the technological changes that might influence the industry. Creative adaptation of technological advancement can suggest possibilities for new plastic products, for improvements in existing plastic products or in manufacturing and marketing techniques. All manufacturers need to have an understanding of both the existing technology and advances and probable future advances that can affect their plastic products and services.

Ecological Factors

Ecology refers to the relationship among human beings and other living things and the air, soil, and water that support them. Threats to life supporting ecology caused principally by human activities in an industrial society are commonly referred to as pollution. Specific concerns about manufacturing and the use of various plastic products in packaging industry include global warming, loss of habitat and bio-diversity as well as air, water and land pollution.

According to UNEP (2005) report, global climate has been changing over the years and it is now evident that humanly activities are accelerating this tremendously. A change in the atmospheric radiation, due in part to ozone depletion, causes global warming. Solar radiation which is normally absorbed in to the atmosphere reaches the earth's surface, heating the soil, water and air, loss of habitat and bio-diversity. Ecologists agree that the extinction of flora and fauna is occurring at a rapid rate and if this pace continues may constitute a global extinction on the scale of those found in fossil records. The earth life forms are dependent on a well-functioning ecosystem and immeasurable advances in disease treatment can be attributed to research involving substances found in plants. As species become extinct, the life support system is irreparably harmed. The primary cause of this extinction is the disturbance of the natural habitat. Acid rain or rain contaminated by carbon dioxide and sulphur dioxide, which can destroy aquatic and plant life is believed to result from plastic waste materials burning. In other words, a health threatening 'thermal blanket' is usually created when the atmosphere traps these gases emitted from smoke stacks in landfill burning plastic wastes. This 'greenhouse effect' can have disastrous consequences, making the climate unpredictable and raising temperatures.

Plastic manufacturers can also dump waste toxins differently into the water bodies, thus causing water pollution. Land pollution is caused by the need to dispose of ever increasing amounts of plastic waste. Routine everyday packaging is a major contributor to the problem. As major contributors to ecological pollution, most plastic manufacturers are now being held responsible for emitting toxic by-products of their current manufacturing process. Managers in plastic manufacturers are required to incorporate ecological concerns into their decision making.

Plastic packaging is the ultimate symbol of the 20th century consumer culture. It protects what people buy and raises their standard of living. In developing countries, 30 % to 50% of food shipments are spoiled because of inadequate packaging and distribution systems. Plastic packaging not only protects goods but also conveys information about their contents and preparation of administration. It plays a vital and growing role in the global economy. At the same time plastic packaging is on the front line. It is the largest and fastest growing contributor to one of the most troubling environmental problem which is plastic waste garbage.

CURRENT SIZE OF PLASTIC PRODUCTION INDUSTRY

By 2014, the value of the world market for plastics packaging materials was estimated at US$ 142 billion and has the potential to reach US$ 190 billion by 2020. The growth forecast for the period 2013 to 2018 is 5.2%, making this type of packaging the fastest growing in the plastics industry. The US is the largest market for plastics packaging, whereas Eastern Europe is the fastest growing market in this type of packaging. One of the areas that utilize the largest volumes of plastic materials in packaging is the beverage manufacturing sector, mainly the soft drinks.

Some of the key players in the plastic packaging industry worldwide are Amcor, ALPLA, Berry Plastics, Graham Packaging, Logoplast, Plastipak, Silgan, among other. Five out of these seven players are US based and two are European based.

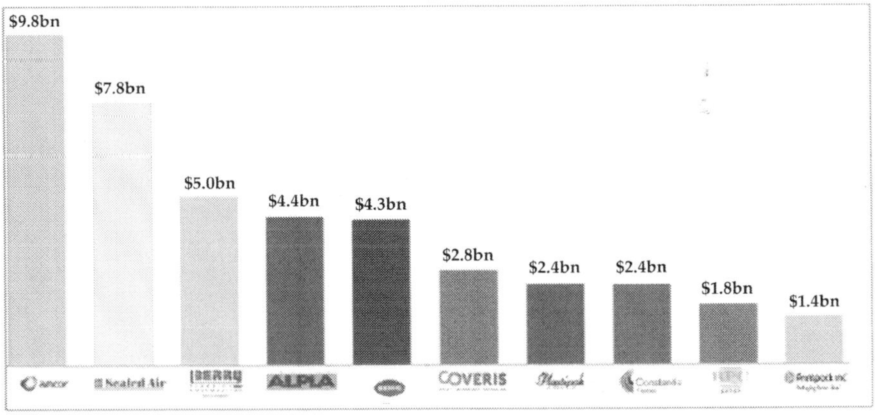

Source:https://www.linkedin.com/pulse/view-top-10-global-plastic-packaging-market-companies-fauquembergue

Figure 3.5: TOP 10 Global Plastic Packaging Market Companies.

Although the growth of the plastic packaging industry in the world is different across regions, there are some factors which influence the overall growth of this industry on a rather long-term basis. Some of these factors include the following;

- Population ageing in the world
- Increase of slammer households
- The demand for products convenience among different consumers
- Enhanced health awareness among different end consumers

- Demand for enhanced branding in an increasingly competitive environment
- Development of novel plastic packaging materials
- Innovation in manufacturing of plastic packaging products
- New plastic packaging regulatory requirements
- Looking at all those factors, the one which is ranked the highest in regards to being important and influential on the plastic industry growth is the 'Health awareness'. The lowest ranked one is the 'Ageing of the world population'.

On the other hand plastic packaging industry is facing some serious challenges. As the macroeconomic environment has been challenging is some countries, the pressure on consumer spending has gone up. For instance, the economy in Europe has been somehow uncertain and the raw material and energy prize inflation has had a negative impact on plastic packaging producers overall.

With the growing concern of environmental issues and the introduction of new environmental legislations, the plastic packaging industry is facing a huge challenge on how to respond to consumers in a way that they keep their faith in this type of packaging solution. The plastic packaging industry is trying to find solutions by reducing the amount of packaging used by light weighting of materials and by using PCR (post-consumer resin) in the packaging (e.g., plastic bottles).

Statistically,TechNavio, a British based market research company, predicts that the global plastic packaging market has grown at a compound annual growth rate of 6.06% from 2014-2019 with a predicted growth of 4.72% for the rigid plastics packaging market. This is compared to other types of packaging such as glass (3.9% compound annual growth rate) and corrugated boxes (3.98% compound annual growth rate).

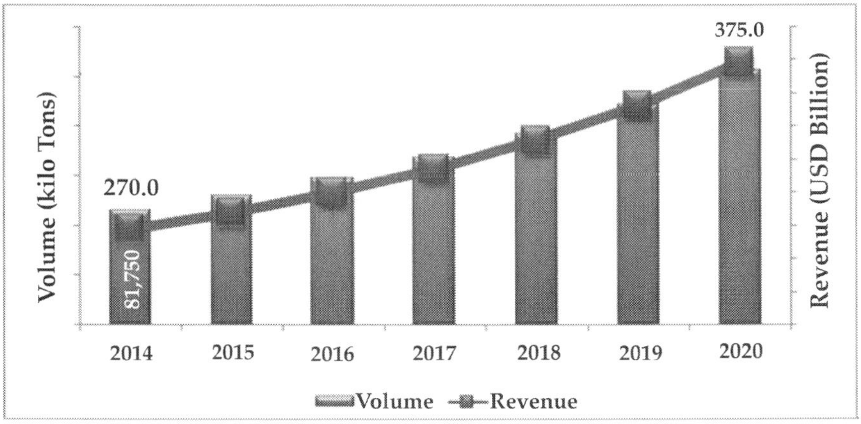

Source:https://www.plasticstoday.com/packaging/global-plastics-packaging-market-hit-375-billion-2020/23800481624973

Figure 3.6: Illustrating the global plastic packaging market.

FACTORS THAT AFFECT THE SUCCESS OF PLASTIC PACKAGING INDUSTRY

There are several factors that may affect the success of a plastic packaging industry in one way or another. Some of these factors include;

- Management of raw material inflation
- The reduction of waste
- Effective capital expenditure
- Operational performance measurement
- Product and customer profitability management
- Innovation and global supply chain management

Management of Raw Material Inflation

This refers to the necessity of companies in the packaging industry to manage the stability of their input raw material costs and pass through as much of the raw material price inflation as possible to the customers. They are typically in a very vulnerable position as their customers are very often powerful consumer goods companies that put a tremendous price pressure on them as they want to be competitive on the market.

Reduction of Waste

Due to increasing environmental concerns, plastic packaging companies are investing in more sophisticated machinery in order to increase their process stability to more effectively manage their waste as well as they are attempting to reduce the material content which is in the plastics industry called 'light- weighting'.

Effective Capital Expenditure

The capital intensiveness is relatively high in plastic packaging industry due to the sophisticated machinery which is required to meet the quality standard the customer requires. Management needs to be experienced in managing the balance between maintaining the existing equipment and investing in new technology in order to be able to be competitive. A beneficial matrix to assess performance in that area is the "spread" between Earnings before Interest, Tax Depreciation, and Amortisation (EBITDA) and capex. In other words "Capital expenditure" is equals to EBITDA minus capex.

Operational Performance Measurement

To achieve continuous operational excellence, it is important to measure the right things. Therefore it is important to have the most meaningful KPI's (key performance indicators) in place. In other words, "Whatever gets measured gets delivered".

Product and Customer Profitability Management

Regarding product and customer profitability measurement, management needs to have the possibility to rank their customers by profitability. Companies that fail to do so risk ending up with a high number of low-margin products which take up the available capacity without making any profit.

In terms of innovation, demographic changes such as the decline of the nuclear family, increase in average age, and increase of single households and increased market share, competition have changed and will continue to change the plastic packaging industry. To respond to those changes, plastic packaging companies have to be innovative and need to come up with new shapes, new materials, more colours, and deliver short-run lengths in an economic way. Additionally they need to be close to the end-market consumers by fostering and nurturing collaborative relationships with the customers, as they are closer to the end-consumers.

The factor of global supply chain management refers to managing the costs involved to produce the plastic packaging. As packaging overall yields a typically low-value product, production site location is a key driver of economic performance. Price decreases as distance between the packaging plant and customers fill line decreases. Some packaging companies are even located in the customer's building and thus feed the packaging (i.e., plastic bottles) directly into the customer's fill line. This approach is called an in-house solution.

Innovation and Global Supply Chain Management

In terms of revenue, plastic packaging market is expected to grow moderately during the forecast period. The major customers of plastic packaging are distributed across various vertical lines such as Food & Beverages (F & B), pharmaceuticals and healthcare, and consumer goods.

4 EFFECTS OF PLASTICS ON MARINE ECOSYSTEM

INTRODUCTION

Industrialisation has led to the discovery of many materials, most of which lack the recovery infrastructure and therefore end up as waste. One of such materials that have been useful in today's industrialisation is plastic, which has become a new form of waste generated within the last 3 to 4 decades. Human garbage now includes various packaging plastics and other synthetics materials. Most of these materials are usually disposed inappropriately and find their way into the marine ecosystems.

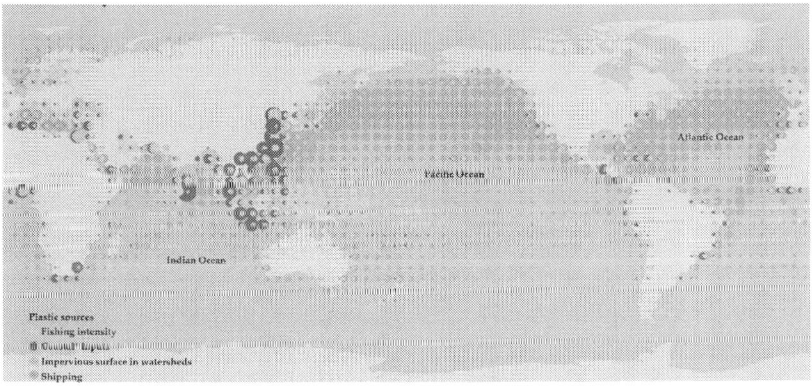

Source:https://ensia.com/photos/plastics-impact-worlds-oceans-outlined-8-maps/

Figure 4.1: The situation of plastic input in marine waters.

According to UNEP (2005), about 80% of all marine debris or litter originate from land, while the remaining 20 % is ocean - based. For example, over the last three decades, the increasing proliferation in the use of synthetic products such as plastics has resulted in a proportionate increase in plastic wastes deposited into the waste stream. Most of the plastics in marine have fatal consequences on the functioning of marine species. Plastics and other marine debris have affected the survival and reproduction of species in most cases causing ingestion and entanglements of aquatic lives.

The properties of plastics which make them so perfect for the manufacturing of other products such as its lightweight, strength, durability, corrosion-resistance with high thermal and electrical insulation also makes them non-biodegradable. Therefore, the continued accumulation of plastic wastes in the marine (onshore and off-shore) poses a serious global problem that is on-going. The flora and fauna of the ocean floor to the surface of the sea, and the whole marine environment is under attack from the effects of plastic accumulation.

Current forms of litter found on most of the beaches have remained largely plastic wastes mostly found on the sea bed and on the surface of sea water forming marine debris. An internationally agreed definition of marine litter, which is adopted by UNEP is "any persistent, manufactured or processed solid material discarded, disposed or abandoned in the marine and coastal environment." An inspection of most beaches in the world reveals the presence of plastic waste seen now as a menace. Globally, the statistics on plastic wastes on the beaches keep rising day by day. Once these plastic materials find their way into the marine ecosystem, they usually become part of the marine litter and may affect the life of many animals in one way or another.

According to UNICEF (2011), the disturbances caused by marine debris, in this case, plastic marine litter to the marine environment can be broadly grouped into three categories. These are ecological, economic and social. First, under the ecological section species such as plants and animals may suffer mortality or sub lethal impacts through entanglement, physical damage and ingestion among others. Also, the swallowing of micro plastics and plastic chemical accumulation may form other potential impacts on the ecological system. Second, plastic marine litters have economic costs to the environment, which includes damage on various vessels, challenges on fishing operations, marine cleaning, and potential challenges to the tourism sector. Third, the social cost of plastic marine litters includes the reduction in aesthetic values of the environment and potential hazards to public safety are some immediate costs.

CHARACTERISTICS OF PLASTIC DEBRIS

Today, most of the waste materials can be found in almost every environment on the globe. Of all the kinds of these waste materials, plastics (especially the packaging and tote plastics) are identified to have the most harmful effects on nature. Plastic waste materials makes up 50- 80% of ocean debris, usually making it the most abundant artificial material, while glass and metal often rank second or third on a global scale. It is estimated that more than 8 million items of plastic are illegally disposed in the oceans every day, while about 13,000 pieces float on every square kilometre of the ocean. In the year 2009, some volunteers collected more than 10 million items of marine debris on beaches of 108 different countries, with the top three items being cigarette butts, plastic bags and food containers. According to another study that was conducted in 2009 concluded that in the OSPAR Maritime Area, the amount of marine debris materials remain high and are not reduced in any way. On average, approximately 70% of the marine debris sinks to the bottom due to fouling processes, while 15% floats and 15% are usually washed ashore. Plastic materials are in general not biodegradable and persist for extended periods of time. In combination with the fact that their demand is rising, the quantities of plastic debris will continue to increase.

Plastics are generally buoyant in sea water, with only a small proportion having a specific gravity greater than that of sea water.Once these items reach the seafloor they will probably be buried and last indefinitely. This can have major environmental consequences; many consider the seafloor as a permanent sink for marine debris. Accumulation patterns of debris on the seafloor are still poorly understood. For instance, it is believed that bottom salinity fronts can act as plastic debris accumulation barriers. Other studies found that most plastic debris usually have the tendency to accumulate in areas with high rates of sedimentation and low rates of water circulation. Organisms settling on floating plastic can increase the density of the items so they sink. Grazing organisms can clean these surfaces from time to time and re-emergence of plastic debris can occur.

Plastic debris on the seafloor can create hard bottoms. These attract sessile organisms and can alter sea floor communities and the composition of natural ecosystems. These blankets of plastic debris can hinder exchange of various gases especially between overlying waters. This is believed to cause various effects on marine species such as anoxia and hypoxia, thereby altering the normal life and functioning of various species.

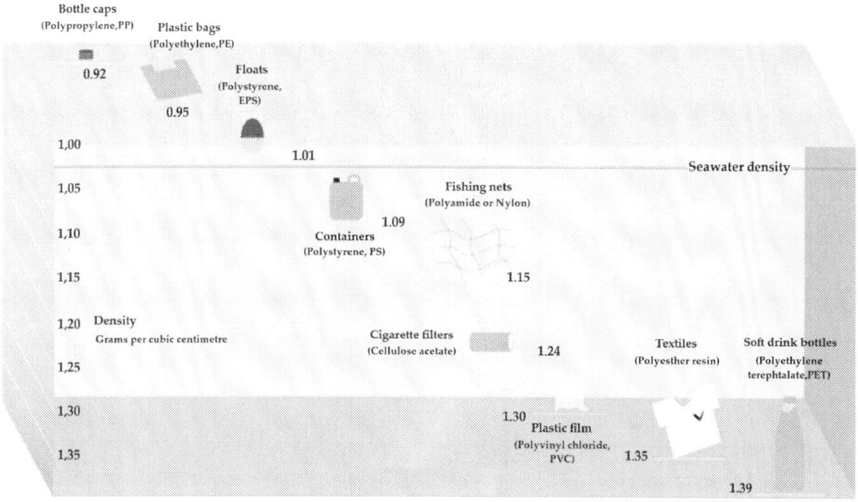

Source:https://ensia.com/photos/plastics-impact-worlds-oceans-outlined-8-maps/

Figure 4.2: Illustrating which plastic float and which sink in the seawaters.

For the last few decades, there has been mass production of plastics, accumulation of plastics in the terrestrial and marine environments. These synthetic organic polymers (plastics) usually act as key contaminants whose effects have gone beyond the visual. Their low cost, longevity and disposable nature have made plastic debris a universal problem. According UNEP (2005) report, what was simply seen initially as an aesthetic problem has become a threat not only to animals and the environment but for humans too. The attractive qualities of plastics make them suitable for the production of numerous products. Unfortunately the same properties are the cause of many environmental problems. Because most plastics are buoyant and persistent in sea water, they have the ability to disperse longer distances than high density particles (such as scrap metals and glasses).

Globally the average size of plastics is decreasing, although the environmental consequences of these small plastics are still largely speculative. According to WHO (2011) report, it is worth noting that over-packaging is a problem as packaging normally sells a product, but because plastics are so persistent in the environment, compared with other discarded material (paper, metal, glass), it is difficult to determine the sources of plastic litter.

An important environmental concern, especially on marine water resources is the issue of plastic littering. This can be overcome through having stricter laws and education programs as littering is mostly described as a behavioral. The existence of plastic litter on beaches and in waterways can have a number of impacts.

- It can decrease the aesthetic value
- It can pose a potential health risk for humans
- Aquatic animals are in danger of getting entangled
- Suffocating or ingesting litter while searching for food
- Pathogenic organisms and toxins can be taken up through the food chain
- Poisoning animals and impacting humans
- Invasive species can be transported
- Can increase economic threats
- Large amounts of cash are spend every year for regular beach clean-ups

Debris can enter the marine environment either accidentally or deliberately.Fishing industry is the main source of marine plastic pollution. This is because it contributes most to ocean pollution by dumping various plastic waste materials including various plastic containers that are used in packaging and storage of fishes once they are caught. Different studies have shown that fishing industries mostly affect shores with adjacent fishing ground.

Beach users are also huge key contributors of plastic pollution in the marine. According to UNICEF (2010) report, it was found that most beaches especially in the peak tourist seasons are usually littered mostly by beach users, but drainage and outfall systems are the main sources in non-peak seasons. Some of the key plastic litters mostly consist of single-use items and plastic packaging containers such as plastic food containers, cans, plastic bottles, and cigarette butts among others.

PLASTIC WASTE ENTRANCE INTO MARINE ECOSYSTEM

There are different route through which different microplastic enter into the marine waters. Primary microplastics (those already in microplastic form) often enter the marine environment via transport by public water systems, direct runoff from industrial processing facilities, and spills from shipping containers; these usually occur in the form of postconsumer facial cleansing scrubbing beads and pellets used in the production of manufactured plastic products. Microplastics from secondary sources (those derived from the degradation of larger discarded plastic pieces) often occur as fragments and filaments shaped from the inconsistent nature of weathering processes.

Large plastics along the beaches are usually exposed to different environmental forces that facilitate their degradation into smaller plastic particles. In particular, direct exposure to sunlight, high oxygen availability, and constant tidal action of beached plastics corrupt their structural and chemical integrity and render them susceptible to fragmentation.Additionally, inadvertent plastic deposition from marine vessels contributes to microplastic pollution; everything from the synthetic fishing line used in small-scale fishing activities to the displaced cargo and litter from large container ships may directly enter the marine system and become subjected to the processes of photodegradation and physical weathering by ocean movement.

The resulting small plastic particulates are then carried into ocean systems and concentrate in the neustonic zone in the upper water column, where they may easily enter the food chain via contamination of or ingestion by marine organisms.Denser microplastics may sink deeper below the surface, where they are more accessible to a greater variety of marine life and trophic levels.Collectively, these microplastic particles are projected to contaminate a majority of the world's oceans at a few fragments per hectare of ocean surface (with greater concentrations in areas of higher anthropogenic pollution and presence).

IMPACTS OF PLASTICS ON MARINE LIFE

The class of contaminants called microplastics is one of the major components of global accumulation of plastic debris. In spite of this, their characterization and identification has occurred only relatively recently within the scientific and environmental communities. Microplastics as ocean contaminants were first characterized by Carpenter and Smith in 1972, who observed large quantities of small plastic particulate floating in the neustonic zone of the North Atlantic Ocean. Frequent observations of microplastics in marine environments were made subsequently, prompting different researchers to attempt a comprehensive categorization of different types of microplastic by type.

In spite of a growing number of studies observing small plastic debris in marine environments, microplastic was not recognized as a distinct class of pollution until 2008, when it was subdivided into macro (>50 mm in diameter), meso (~50-5 mm), and micro (<5 mm) pieces. This formal designation (along with a host of subsequent studies examining its toxic effects on marine habitat and biota) rapidly hastened the examination of microplastic as an environmental pollutant and remains its defining characteristic.

Hazards for Marine Environment and Biota

There are serious environmental hazards experienced by marine ecosystem. Due to their light weight, plastics are often subject to the movements of ocean currents and may be transported thousands of miles away from their entrance point, resulting in widespread contamination globally as well as significant changes in marine substrate locally. Therefore, patterns of plastic contamination may increase the population density and mobility of species in areas where they were not previously, possibly disrupting predation and resource allocation in localized food webs. Furthermore, photodegradation of microplastics may have significant environmental consequences through the release of primary constituent chemicals (used to enhance plastic quality and durability), which may alter ocean chemistry and render conditions unsuitable for resident marine biota.

Plastics also impact marine organisms directly through ingestion and subsequent blockage and contamination of internal systems. Plastics may remain suspended in the upper water column for long periods of time, where they may be directly ingested by marine predators or inadvertently consumed by filter-feeding organisms. Furthermore, most secondary microplastics assume a variety of different shapes and colours, making them attractive food items for marine predators such as fish and seabirds.

There evidences of plastics consumption by vertebrates and invertebrates alike, particularly predator species that pursue prey along the surface and filter-feeders that filter nutrients at lower depths. For instance a number of scientists have found evidence of plastic in the stomachs of different species of forage fish studied from Washington State, confirming the presence of plastic via ingestion, a process that potentially disrupts digestive function.

Similarly, a study conducted by University of Nairobi, Kenya found higher levels of brightly-colored high-density polyethylene in northern fulmar stomachs than any other plastic type, indicating its attractiveness as a target prey item. Microplastic also has been identified in a number of filter-feeding organisms, namely mussels and oysters.

Some marine animals eat marine debris either intentionally or unintentionally, when this debris is mistaken for food. Its effects on the animal will depend on the shape, size and volume of plastics ingested. In fewer cases, plastics can go through the digestive tract without causing any harm to the marine animal, but in most cases, the plastic becomes lodged in the throat or digestive system leading to strangulation, starvation, malnutrition and eventual death.

Again, ingested plastics cannot be digested and may lead to a sign of fullness of the stomach which blocks the intake of extra food leading to starvation and malnutrition. Further, plastics make the animal full leading to restricted ability to swim and dive. Though cetaceans, turtles and birds have been identified as being the most affected by ingested plastic marine debris worldwide, fish of all types have also been identified to ingest plastic marine debris.

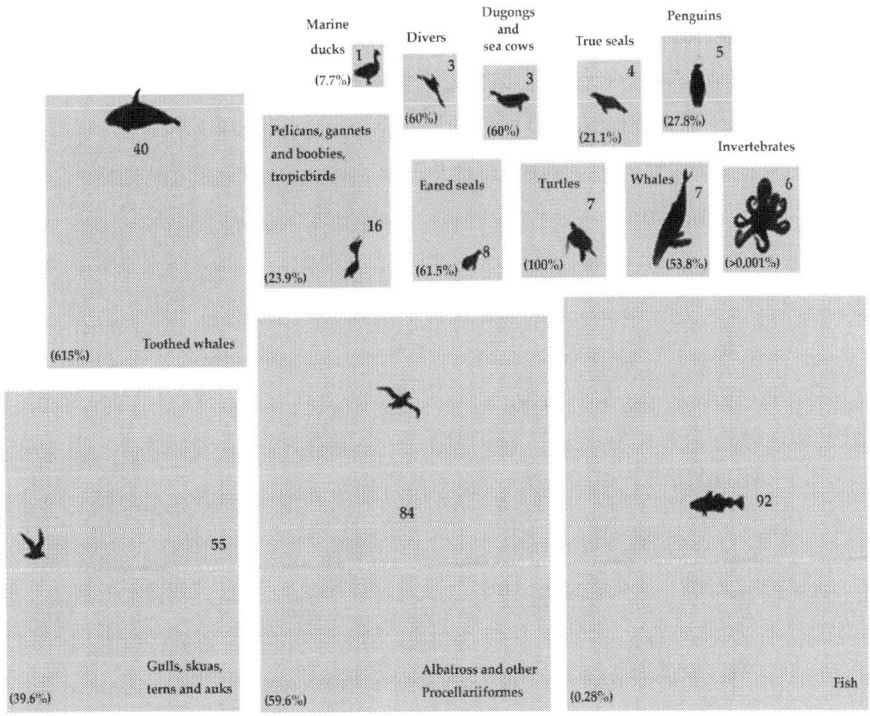

Source:https://ensia.com/photos/plastics-impact-worlds-oceans-outlined-8-maps/

Figure 4.3: Illustrating the total of marine species that are documented in records that suffer from plastic waste ingestion in marine water resources.

In addition, plastic debris ingested by marine animals, have the potential of transferring hazardous chemicals in the plastics into the animal's body which could potentially have lethal consequences on the animal.

Recently-discovered chemical hazards associated with plastic ingestion pose a more subtle but just as potent threat to organismal and ecosystem health. These hazards include the leaching of primary chemical constituents into biotic tissue post-digestion and accumulation of adsorbed environmental chemicals from the surrounding ocean matrix in biotic tissue and organ systems. These latter two mechanisms (leaching of primary raw chemicals from "clean" microplastics and deposition of pollutant chemicals by contaminated microplastics) are the two most implicated but often overlooked pathways of chemical transference.

Bio-accumulative Plastic Toxins

Direct leaching of existing primary chemical constituents into biotic tissue is an inherent risk associated with microplastic consumption. If microplastics remain and accumulate within internal systems for long periods of time, enzymatic activity may induce chemical deposition of constituents from within the microplastic chemical structure into tissues. In contrast, secondary leaching of adsorbed chemicals only recently has been characterized in the wake of microplastics identification as a potent vector of environmental pollutants. Preliminary research on the chemical-adsorbing properties of microplastics has identified them as effective transport mechanisms for environmental toxins.

Given their physical durability, chemical stability, and hydrophobicity due to their nonpolar composition, microplastics are ideal carriers of persistent bioaccumulative toxins (PBTs), anthropogenically-derived environmental toxins that exist in the environment. These uncombined monomers and persistent organic pollutants adsorb onto plastic via a chemical process called partitioning, in which the toxins effectively coat the surface of microplastic fragments. Like plastic particulates, PBTs exhibit strong hydrophobicity in water, facilitating their accumulation or "partitioning" from the surrounding seawater onto the microplastic particles. Additionally, common microplastic shapes (flat fragments, spherical beads, filaments) have an increased surface area to volume ratio that optimizes the accumulation of environmental toxins. Recent studies have shown their effectiveness as chemical sinks when pollutant levels in microplastics were found to be comparable to those in sediment concentrations of the same chemical compounds, effectively implicating microplastic particles as PBT vectors. Thus, there is strong evidence pointing to microplastic particles as conduits for PBTs, partitioning chemicals from the water and carrying them through the ocean in small but highly-concentrated doses.

Ingestion of contaminated microplastics by marine organisms may deposit high concentrations of toxic and bioaccumulative chemicals into biotic tissue and the larger food web. There is evidence for the bioavailability of polycyclic aromatic hydrocarbons in microplastic particles in the marine lugworm digestive tract, discovering that gut surfactants of benthic organisms concentrate ingested PBTs. Chemical transference from microplastic particles also has been identified in larger marine predator species such as the Great and Flesh-footed shearwater, indicating the ecological extent of microplastic contamination. Chemical transference of PBTs by microplastic ingestion is of particular and increasing concern to the health of marine species and the ecological integrity of marine ecosystems; chemical leaching of environmental pollutants within the digestive system may be absorbed in biotic tissue and induce adverse and lethal physiological effects in marine organisms.

One class of PBTs, polychlorinated biphenyls (PCBs) has been particularly implicated as a harmful marine contaminant and bioaccumulant in organisms. Indeed, many studies on the partitioning chemistry of microplastics have observed high concentrations of environmentally-adsorbed PCBs within microplastic fragments.

Polychlorinated biphenyls are nonpolar and exhibit strong hydrophobicity in water, facilitating their easy adsorption onto microplastic particles as shown in the figure below.

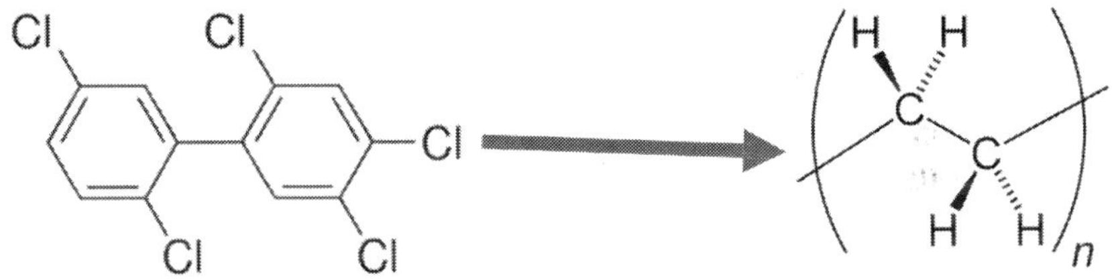

Source: https://en.wikipedia.org/wiki/Polychlorinated_biphenyl

Figure 4.4: Polychlorinatedbiphenyladsorbsontopolyethylenefragment.

Furthermore, PCBs have extremely high toxicity and durability, allowing them to remain in and move through the water matrix for long periods of time without being degraded. This high mobility in water allows PCBs to bioaccumulate easily within marine organisms, infiltrating local food webs and contaminating higher trophic levels via biomagnification, a process by which toxins in prey tissue are transferred via ingestion and concentrate within dominant predator species. Ingestion and accumulation of environmentally-contaminated microplastics by secondary consumers (such as fish) therefore has a disproportional effect on tertiary and quaternary consumers whose diet consists primarily of lower trophic-level species.

Human Health Concerns

While plastic in its inert form is not especially dangerous to humans, its degradation in the marine environment poses significant potential hazards to human health. Oceanic weathering of plastic via photodegradation and mechanical abrasion may release constituent monomers, additives, and chemicals present within microplastic particles, contaminating marine environments used by humans for food and recreation. Perhaps more significantly, microplastics may accumulate within lower- trophic level organisms, biomagnified through the food chain, and deposit high concentrations of plastic and adsorbed toxins in upper trophic consumers. As all trophic-level organisms may then be contaminated by microplastic to some degree, any organisms upon which human populations depend for food may be contaminated, from primary consumers (i.e. mussels) to secondary and tertiary predators (i.e. tuna). Subsequent ingestion of contaminated fish tissue may induce slight physiological effects in human systems; however, it is important to note that these concentrations are often too low to detect after seafood processing and preparation and effects are often negligible or non-existent.

Additionally, microplastics often accumulate within the organism's gut, which is not a popular food item for human consumption; therefore, transference of microplastic particles from seafood to humans largely depends on translocation of microplastics through the organism's system to target tissues of consumption. However, some organisms are an exception; the whole-body consumption of filter-feeding molluscs is a potential concern, especially given the rapid growth of the mussel and clam industries and recent identification of microplastic accumulation in mussels and oysters. Furthermore, internal leaching of toxins from gut-accumulated microplastic has been documented in a number of marine species and may pose a health concern for human consumption on contaminated tissues. More studies tracing the translocation of ingested microplastic to specific systems and tissues should be conducted to form a greater understanding of how plastic accumulates and moves within marine biota.

There are other different impacts of plastics on the marine life. These impacts can be classified under four major categories.

Entanglement: Entanglement in marine debris is one of the causes of widespread animal suffering, leading to the deaths of about a hundred to thousands of marine animals and birds each year. There are several species that may suffer from entanglement such as marine mammals, seabirds and sea turtles.

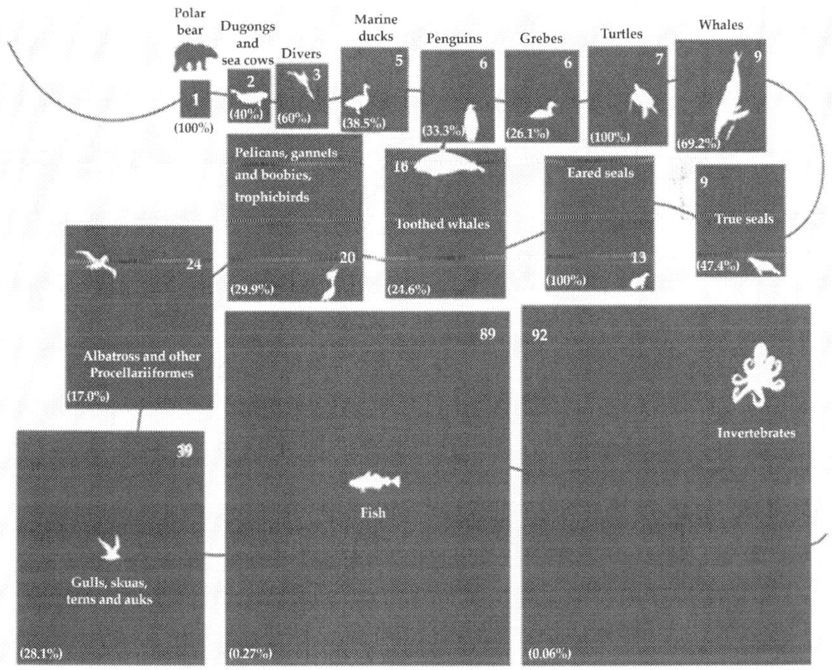

Source:https://ensia.com/photos/plastics-impact-worlds-oceans-outlined-8-maps/

Figure 4.5: Illustrating the total number of marine species with documented records of entanglement in plastic waste materials in marine water resources.

Entanglement takes place when fishing gears made of plastic material are intentionally or unintentionally abandoned such as monofilament line, nets and ropes, as well as other types of ropes, nets and plastic packaging from consumer goods. Old or worn out fishing gear discarded in the oceans may continue trapping marine organisms such as fish and crustaceans as its gets carried away by the current and end up killing such trapped species. The manner in which marine animals are entangled and the type of debris involved differs by species; basically, the animal's body shape and behaviours may impact on the nature of entanglement.

Entanglement may affect marine life differently. For example, plastic debris may restrict the ability of the animal to feed, leading to starvation. The additional materials trapped may exert undue weight over marine organisms whiles hindering its movement and inability to escape impending predators ultimately strangling them to death. Rope and ligatures may cause amputation or severe wounds that leave sores and infections on marine organisms. Loops and hooks may cause deep cuts into the skin muscle. It has been noted that the non-biodegradable character of plastics gives it the potential to entangle more sea animals after the first is dead and decomposes.

Invasion: The impact of invasive species especially on various indigenous marine species has been identified as a key reason why the biodiversity is diminishing day by day globally. For instance, over the last few decades opportunities have increased for invasive organisms, especially fouling organisms, to travel long distances. Natural transport for marine organisms is being supplemented by plastic debris and shipping activities. For this reason therefore, shipping activities have been identified as the main source for introducing species to other locations. Indeed, plastic materials have doubled most of these opportunities for organism dispersal in the tropics but humans have actually aggravated the dispersals even more especially at sub-polar latitudes. Studies have shown that most abundant species that are usually dispersed are species with a cosmopolitan distribution. It is believed that buoyant plastic debris can be a potential vector for micro-algal dispersal.

Introduction or spread of alien species: Anthropogenic activities have led to the movement of some species into regions where previously, they were non-natives. The movement of one species from one habitant into another is usually referred to as invasion, and these invasions may have devastating consequences for the ecosystems concerned. Such movements may result in the transfer of organisms or chemicals on organic materials that suspend in water within the marine environment such as seeds, tress, and leaves. Research studies have shown that plastic litter floating on the surface of the ocean have increased the number of rafts that transport environmental pollutants.

Damage to coral reefs: Old and abandoned fishing gears, especially lobster trap, are known to be a source of destruction to coral reefs. Ligatures and lines become entangled with coral reefs and eventually break off at the point of entanglement. Once the debris is freed, the process continues until the debris becomes part of the whole structure of the coral reefs.

According to a recent research study that was conducted on the North western Hawaiian Islands to determine the effect of plastics on corals revealed that these plastics cause damage and death to different species include the corals and sponges.

SOLUTIONS TO THE PLASTIC MARINE DEBRIS MENACE

It is evident that plastic waste pollution in the marine environment has had some devastating effects on the activities that are undertaken on the ocean and sea waters.

Measures to address marine plastic waste pollution can be categorized into three types, preventative, mitigation and curative. Preventative measures to reduce marine debris are not only most effective to solve the problem; they are also less costly. Once the plastic wastes have reached the ocean, collection and the necessary separation to recycle and incinerate the plastic waste materials properly are expensive and can harm the environment. Therefore, preventative measures against marine plastic pollution need to be focused on tourism as well as on the shipping industry.

Preventive Measures

Proper enforcement is often difficult and little data is available on the effectiveness of these regulations. The responsible bodies for coastal protection and other local municipalities should work together with coastal communities to remove beached plastic wastes. In the absence of any meaningful preventive intervention from the responsible bodies (such as fisheries and aquaculture development office, municipal assembly, and other governmental agencies) fishermen have no other means but will have to cope with the situation until some sort of intervention preventive measures are initiated by the municipality. Understanding the coping mechanisms of the fishermen through the use of indigenous technology as well as borrowed technology is a huge step towards addressing the problem.

Apart from the above measures, there are other several preventive measures that can be taken in order to reduce plastics pollution in marine waters. Globally, many preventive intervention programs are being put in place to help manage marine pollution. The UNEP and US EPA among many others known international organizations have put in preventive measures to help curb this global problem. For instance, many organisations have come up with preventive measures for reducing or halting the indiscriminate disposal of litter especially that of plastics and other synthetic materials into oceans and seas around the world. Many of these organisational preventive initiatives can be grouped into global, international and national initiatives. According to a report given by Convention on Biological Diversity in 2012, it was stated that "marine plastic debris could be reduced through eco-labelling, green procurement, extended producer responsibility; deposit return progress; fees charged on single use plastic bags; education to promote the viewing of wastes as a resource for recycling; engaging with corporations and industry associations on sustainability; encouraging reuse and reduction through green chemical; encouraging better product and packaging design and lastly, supporting marine debris awareness."

The international Convention for the Prevention of Pollution from Ships (MARPOL) is also one of such global negotiations working around the clock to ensure the prevention of plastics and other synthetics from entering the sea. MARPOL's 1988 legislation aimed at preventing ships from disposing their garbage over board. This legislation restricts the dumping of packaging plastics over board and also urges all member states to maintain garbage receptors at the various sea ports to discourage ships from disposing wastes over board.

The Global Program for Action (GPA) for the protection of ocean from land base activities is also a UNEP collaborated program with other intergovernmental oceanographic commissions of the United Nations Educational Scientific and Cultural Organisation (UNESCO-IOC) on strategies on monitoring marine litter from entering the sea from land based sources since it forms about 80% of all marine debris.

Different privately owned companies across the world in charge of waste management have also adopted some preventive measures to help curb the ever increasing plastic waste problem that end up into the oceans and seas. According to UNICEF (2010) report, other form of preventive measures are being given to the general public on plastic waste management and more importantly to prevent plastics from entering into oceans and seas, since most beaches around the world also serve as dump sites. These dumps inevitably have some consequences on various activities that happen on ocean and seas water.

Source: http://www.huffingtonpost.com/jeb-harrison/sea-of-japan-becoming-a-dumping-ground-for-trash-from-china-and-south-korea_b_8005772.html

Figure 4.6: Illustrating a typical dumpsite along the beach.

It is believed that if bins are provided at the beaches, they will use it for its purpose. Arresting people who throw rubbish at the beaches will also serve as deterrent to others. Additionally, laws should be enacted on plastic waste disposal. It is also worth noting that fisherman associations should be strengthened to serve as a unified force in dealing with the plastic wastes problem and any other problem that may confront the fishing occupation.

Another preventive measure that can be taken at community and national levels entails educating the general public population about the effects of plastics waste pollution in sea and ocean water resources. This education should include discussion of the threats of marine plastic waste debris on nature and human health. To effectively educate people who are working in this factor contributing to marine plastic waste, education programs for coastal communities could be provided at the local schools or education centres. Similar education programs could also be provided for tourists. Furthermore, it is recommended to involve NGOs. Education of children is also very significant. Children are more likely to develop environment friendly behaviours if they are confronted with the threats resulting from pollution early. Furthermore, early on confrontations are important, as the next generation will inherit the problem of marine plastic waste pollution if measures for plastic waste reductions are not taken. It is also believed that if fishermen are educated on the dangers that plastic materials pose to the fish and other marine species on which their livelihoods depends, most of them will not dispose of plastics into water bodies.

Curative Measures

Although beach clean-ups can only be a temporary solution since a large proportion of the marine plastic wastes returns to the beach, in most places around the world, coastal beaches are not frequently cleaned. In any case, a beach clean-up should strongly be recommended. Here the density of plastic waste materials is especially high and visual effects could harm the tourist industry in addition to the negative environmental consequences. For the clean-up operation, ideally, the whole community around coastal beaches (including schools and the University Centres) and surrounding villages would participate in the clean-up operation. The beach clean-up could include an analysis of the quantity and quality of plastic waste items accomplished. Some of the packaging plastic container wastes might even be of value for arts or decoration.

Other curative measures to reduce plastic waste pollution in marine ecosystems include, plastics products especially the plastic carrier bags should be banned and replaced with paper bags. Security should also be provided and strengthened around the shore to prevent people from dumping refuse haphazardly.

Source:https://ensia.com/photos/plastics-impact-worlds-oceans-outlined-8-maps/

Figure 4.7: Illustrating Key actions different countries around the world are taking in order to combat plastic littering in marine water resources.

Many organizations have researched into measures to take in order to reduce the amount of plastic waste that enters oceans and seas. According to World Health Organization (2005) report, the first step to addressing the problem of marine litter is by knowing the sources of such litter. It has been established that about 80% of all marine plastic litters originates from land base sources. These litters end up in the environment through intentional and unintentional means. Therefore, proper plastic waste management around beaches, water bodies as well as domestic and industrial wastes is crucial.

Several institutions are required to research, formulate and co-ordinate all the activities on marine waters. For instance, countries can come up with necessary commissions and agencies mandated to regulate and manage the utilization of the marine water resources and to coordinate the policies in relation to them. Other curative actions that such commissions can embrace in reducing plastic pollution in marine environment include.

Preparing and keeping under continual review, plans for the management and development of marine waters under the jurisdiction of specific countries

- Establishing priorities for the utilization of marine water resources which will provide the greatest benefit to a country
- Ensuring that proper conservation of the marine water resources through the prevention of plastic pollution
- Striving to minimise as far as practicable various plastic waste materials that find their way into oceans and sea water resources
- Monitoring, controlling and surveillance of marine waters
- Promoting co-operation among local fishermen and advance development of artisanal fishing activities in marine water resources
- Promoting sub regional, regional and international co-operation in management of marine water resources
- Carrying out research and surveillance work for the assessment of marine water resources

Consulting with the competent authority and establishing requirement for manning various vessels and boats in order to avoid dumping of plastic waste material into the oceans and sea water resources

Mitigation Measures

Marine plastic wastes pollution has only been a problem since the invention of plastics. Earlier, tourists and fishermen used gear made from natural and biodegradable material such as cotton.

The existence materials and biodegradable gear on the market shows that it is possible to replace plastic material with biodegradable material with similar properties and with little additional cost. Thus, the government should implement better incentives for the development of biodegradable materials suitable for the environment. If suitable materials are available should be enforced, starting with materials types that are frequently lost at sea and/or shoreline, and further expanding to enforcement of changes to all types of plastic materials. The following are some of key mitigation initiatives that can be taken to curb plastics waste pollution in marine water resources.

Using alternatives: Alternatives to the plastic packaging and carrier materials are available in the market due to the environmental issues surrounding plastics. These alternatives serve the same purpose. The main difference between the initial plastic materials and these alternatives is the material composition. The alternatives are still meant for day to day use and maintain the same general convenience of a plastic packaging material. For instance, BioBag® is one of the important manufacturers that produce 100-percent biodegradable and 100-percent compostable plastic bags. These plastic bags are usually made from a material referred to as Mater-Bi. The company manufactures the bags with starches containing no genetically modified organisms, a biodegradable polymer, and other various renewable resources. The company offers a variety of compostable and recyclable bags including dog bags, cat litterbags, lawn and leaf bags, kitchen bags, and shopping bags.

Source:http://www.biomasspackaging.com/brands/biobag/

Figure 4.8: Biobag packaging alternatives for the plastics.

All of the bags conform to the Biodegradable Products Institute, ASTM D6400 specification for composting in municipal and industrial composting facilities. The company offers the BioShop bag to retailers around the United States in three sizes and specialty custom sizes. In addition, they offer logo imprinting with soy-based colours. The company touts that whether an individual incinerates or composts the BioShop bag, it still has a relatively smaller impact on climate change in comparison to plastic.

Trellis Earth Products Inc. is another company that sells biodegradable alternative plastic bags, however their bags do not conform to the ASTM 6400 standard for compostability. They manufacture a bag made of high quality polymers, starches, and other ingredients. The company claims their bags will naturally biodegrade in a landfill with no toxic residues.

Recently, Whole Foods Market ® announced that they would stop providing customers with plastic bags. Instead the company provides customers with 100-percent recycled paper bags and the option to purchase reusable bags for a discount. However, the Whole Foods paper bags are only 40-percent post-consumer waste and 60-post industrial waste. In contrast, Duro Bag Manufacturing Company sells bags that are 100% post-consumer fibre content. These bags are more favorable alternative to the Whole Foods paper bags because they use more post-consumer recycled waste.

There are a variety of reasons why consumer adoption of these alternatives is not widespread. The alternatives listed above and available on the market today are significantly more expensive than various single-use plastic packaging and carrier bags. The bio-plastic alternative sells for around 7 to 8-cents. Trellis Earth biodegradable bags sell for a premium of 9.2-cents per bag. Paper bags sell for about 4-cents and the classic plastic shopping bag costs approximately 1 to 2- cents. Retailers and consumers are less likely to adopt a bag that is more costly. The main reason the adoption rate on these bags is low is because retailers do not make these bags available to customers.

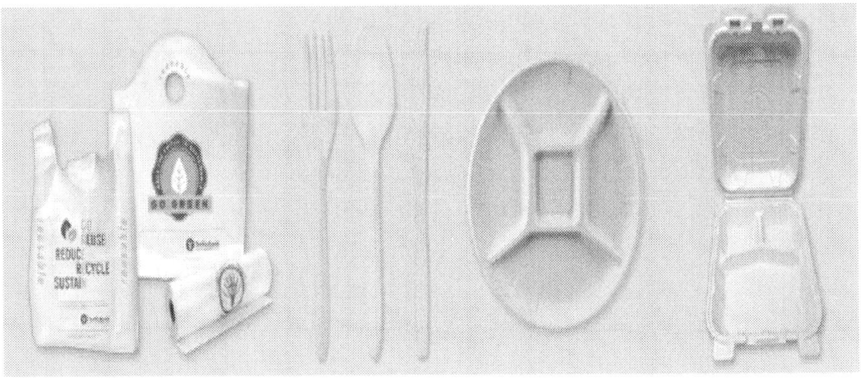

Source:http://trellisbioplastic.com/products-services/bioplastic-disposables/

Figure 4.9: Some of the key examples of disposable plastic packaging alternatives made by Trellis Earth.

Biodegradable and compostable bags require an extensive industrial or municipal composting infrastructure in order to properly breakdown. Unfortunately, the United States does not have an extensive composting facility infrastructure. In addition, these bags will not break down in at-home composting bins; they must breakdown in industrial conditions. Another option is to recycle these bags; however recyclers currently do not have the capacity to recycle bio-plastics.

Consumption of these alternatives would still cause consumers to dispose roughly 326 bags per year. The alternatives still follow the single-use concept, thus consumer consumption patterns are unlikely to change despite the change to more environmentally friendly materials. For bio-plastic bags to replace the tradition plastic bags, it would take significantly more corn, water, and cultivated land. It is worth noting that bio-plastics may contaminate the recycle waste stream, rendering batches of recycled plastic useless. In addition, the use of food products such as corn, sugar cane, and wheat in the production of bio-plastics may increase the commodity prices of these foods.

The paper-bag alternative is also inefficient. Numerous articles reveal that the production of paper bags is significantly more energy intensive than plastic bags. In the production process, these bags emit 70-percent more greenhouse gases than plastic bags.

Authority innovation decision is an influential motivator that instigates adoption and compliance. It has been proven that the effectiveness of authority decisions and policies on innovations are significant to adoption of plastic alternatives. For example, local government taxes and bans placed on initial consumables may prompt individuals to switch to alternative and substitute products. For a drastic transformation to occur in consumption patterns, governments and other authoritative bodies should look at a policy as a tool to promote change. Consumer demand drives production and consumption. A major roadblock for sustainable supply chains continues to be the link to the consumer, where consumers are the decision makers. Consumer actions inevitably determine the fate of the product; whether its consumption is reduced, reused, recycled, or disposed of in a landfill.

Plastics, as consumables, pollute the environment, negatively affect health, drain valuable finite resources, and harm animals. Therefore, consumers should integrate plastic alternative more readily into their lifestyles. Through reducing consumption and reusing alternatives, consumers will help enable the success of sustainable supply chains.

Substitutes for the plastic: Numerous reusable bag alternatives are available to consumers at a variety of retail stores. These bags serve the same general purpose as the initial product and alternatives; however instead of following a single-use concept, these bags are durable enough to be used several times before disposal. These bags have the potential to leave a smaller environmental footprint and are made from a variety of materials including cotton, hemp, bamboo, recycled plastic, compostable plastic, among many other alternatives.

Reusablebags.com is a helpful resource for obtaining sturdy and reliable reusable bags. The company only sells from vendors it approves. The website offers ultra-compact bags, string bags, heavy duty bags, thermal and insulated bags, and printed bags in a variety of materials including recycled cotton and organic hemp. These bags have a wide price range and are typically biodegradable in a landfill at the end of their use. ACME Bags™ is a notable manufacturer of a large array of reusable bags developed by ReusableBags.com. One bag it manufactures is a dual-handled and is made from hemp, a material that is stronger and more durable than cotton; resistant to insect enemies, and 100-percent biodegradable. The company provides a lifetime warranty with all of their bags; thus, the consumer never has to dispose the bag in a landfill. A unique reusable substitute plastic product is one that is made from recycled materials.

Designboom® is another outlet that sells handmade plastic substitute products. They are made from recycled vinyl Bollywood billboards that would otherwise be discarded. ChicoBag™ manufacturers the original rePETe™ reusable bag, made from 99-percent recycled content and 73-percent post-consumer recycled materials. The bags are durable and machine washable for multiple uses. Global Goods Partners is a website that sells a variety of products that benefit women in developing nations around the world. Women in developing nations hand make these bags from recycled materials such as rice bags and laundry detergent pouches.

Source:http://www.bgreentoday.com/green-living-gifts-amp-more/chicobags-reusable-shopping-bags-varieties

Figure 4.10: Key examples of ChicoBags Reusable packaging Bags.

Accustomed to the widespread availability of free plastic bags, a majority of American consumers do not use reusable bags, despite their obvious benefits on the environment. Reusable bags are more expensive than plastic bags and bio-plastic bags, with costs ranging anywhere from \$3.00 to \$45.00.

Retailers provide plastic carriers for no charge. As a result, most consumers have no incentive to switch to reusable bags. Reusable bags also do not have the same convenience as plastic carriers available at stores. Consumers must bring reusable carrier bags from home to the store. The environmental footprint that a reusable bag leaves is significantly less than the initial and alternative options. It has been shown that an individual will dispose less than one reusable carrier bag per year. Brands such as ACME Bags™ have a lifetime warranty, so the consumer impact on landfills would be nothing each year. Many of the bags available on ReusableBags.com are 100-percent biodegradable because they are made from 100-percent natural plant fibres such as cotton and hemp. Cotton rags can biodegrade in 1-5 months unlike plastics that photo-degrade in about 1,000 years.

According to the American Chemistry Council, the average American uses 660 plastic bags a year. The production of these bags consumes 336 mega joules of energy. The production of four reusable cotton or polyester totes uses 140 mega joules or 170 mega joules, respectively. Using four reusable bags consumes 50-percent less energy than single-use plastic bags. In addition, recycled cotton and plastic reusable bags save more energy because the fibre reprocessing is less energy intensive. Used cotton textiles take only 2.6-percent of the energy involved in making new cotton textiles.

Variables affecting adoption of substitute innovation: of the two scenarios, the substitute is the most favourable innovation because it is relatively environmentally friendly and resource efficient. Although the substitute leaves the smallest environmental footprint, consumers are not embracing reusable packaging and carrier bags for regular use in grocery and retail stores. Five perceived characteristics of reusable bags and the innovation-decision process heavily influences the slow rate of adoption of this substitute.

For the consumer, the relative advantage of reusable bags is lower than plastic bags in an economic perspective. Plastic bags are cheaper than other alternatives for retailers. Thus, retailers provide plastic bags free of charge for consumers. Reusable carrier bags are significantly more expensive for consumers, but elimination of plastic bags can save retailers around the United States to save over $4 billion annually. In order for consumers to consider adopting the innovation of biodegradable, compostable and recyclable reusable bags, the economics of the adoption must be more favourable. Incentives can have a profound effect on the adoption of reusable bags by rewarding consumers for a desirable behaviour change. For example, Whole Foods Market offers its customers $.05 to $.10 cents off of the bill for each reusable bag used. There must be some type of relative advantage available to consumers for this innovation to be widely adopted. The compatibility of the reusable bag to sociocultural values and beliefs, previously introduced ideas, and the consumers' need for innovation is an important indicator of the products adoptability. A majority of Americans do not understand the damage that plastic bags have on marine and animal life. Based on secondary research, current sociocultural values and beliefs seem to support the use of plastic as a consumable. As a result, reusable plastic bags are not compatible with current societal beliefs. According to UNICEF (2013) report, "potential adopters may not recognize that they have a need for an innovation until they become aware of the new idea or its consequences". Perhaps if more individuals were educated on the negative consequences of plastic bag usage, the adoption of reusable bags would be more likely.

For this innovation, complexity is not as important of a variable as compatibility. The recyclable, compostable, and biodegradable reusable bag is a simple innovation that is generally not complex to adopt. The reusable bag is more of an inconvenience to Americans, because consumers use free plastic bags. The suitability of the reusable bag is relatively easy. If consumers want to try the adoption of reusing bags, they can bring any type of carrier bag with them to grocery and retail stores. The observation component of the reusable bag is important for its adoption. This innovation is particularly easy to observe and communicate to other people.

The type of innovation decision drives the adoption rate of the reusable bag. Very few retailers and governments have taken a stance on the detrimental consumption of plastic bags. Consumers that adopt the reusable bag innovation are making an optional and independent decision to do so. The authority innovation-decision is particular influential on adoption rates. For example, retailers such as Costco and Sam's Club do not provide their customers with any type of bagging to carry groceries.

As a result, customers carry goods individually or bring in reusable bags. For instance, in Ireland, the government enacted a 15-cent ($0.20

U.S.) tax on plastic bags. The result of this authority innovation decision is a 95-percent reduction in the use of plastic bags. Instead most inhabitants of Ireland carry around reusable bags. Although the amount of tax is widely debated, many experts agree that a tax can effectively deter consumers from plastic bags and shift them to adopting reusable bags.

Environmentally behaviour change: Certain environmental problem can be solved using different approaches. For instance, behaviour change is one way to improve environmental problems. According to UNEP (2010) viewpoint, environmentally responsible behaviour involves exploring how people currently act towards the environment in different settings, as well as a way to inform behaviour through educational programs and outreach that target sustainable action. While behaviour is not the only way to address environmental problems, like marine plastic waste pollution, small changes in a person's actions are steps in the positive direction. Actions to mitigate the amount of plastic trashes entering the marine environment have yet to be studied in a comprehensive setting using theoretical frameworks. For this reason, there is the need to explore how environmental beliefs and attitudes and socio-demographic variables align with one's environmentally responsible behaviour specifically related to plastic waste in marine water resources.

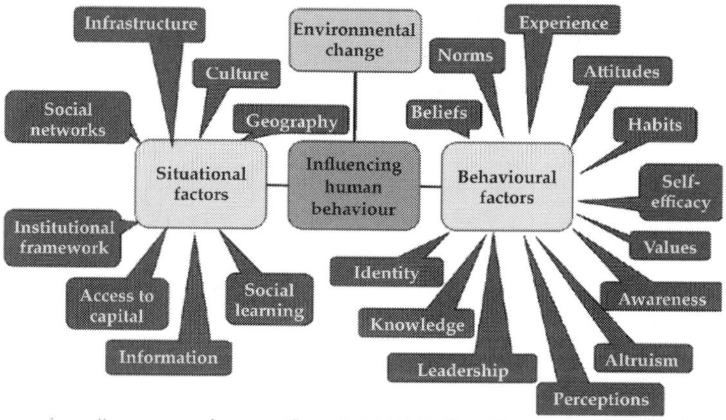

Source:https://www.researchgate.net/figure/258169307_fig1_Figure-2-Situational-and-behavioural-factors-influencing-environmental-behaviour-Defra

Figure 4.11: Key factors that influence environmental behaviours towards mitigating plastic waste pollution in marine water resources.

Additionally, outdoor recreation activities have been shown to influence how a person feels and behaves toward the environment. Participation in recreational activities is thought to enhance environmental awareness and appreciation for one's surroundings. Marine recreation attracts a large number of people. Therefore, in order to understand how recreation influences behaviour, recreationists' behaviour can be studied as a first step to minimize the amount of plastic waste materials introduced into the marine water resources around the world.

Plastic waste materials have actually become a prevalent threat to the marine water ecosystem and this has attracted attention worldwide. Debris of all sizes has become a pervasive global issue due to its ability to travel far distances across the oceans. Because of this, assessing global accumulation of trash in the ocean is quite challenging. Conducting beach surveys is one of the methods that are widely used to evaluate the amount of plastic waste materials that accumulate in marine waters resources.

For instance, according to surveys conducted in Alaska found a reduction in the amount of plastic waste debris from fishing gear and ships has reduced by 60 percent. This example sheds lights on how international intervention can be effective to provide support for the appearance of stabilization in global quantification of plastic debris in marine waters. However, this intervention only addresses the dumping of waste by maritime activities. Another major source of plastic debris is through the deliberate or accidental introduction of materials into the ocean from land. Examples include runoff from storm drains or waterways and trash left by beach-goers. It has been more difficult to control the land-based sources of plastic debris because of the challenges in enforcing regulations against plastic littering laws. For this reason, it is hard to accurately predict how much trash is currently polluting marine waters. Despite the many research studies that show a stabilization of plastic waste materials entering marine waters, it has been noted that the quantity of plastic trash being collected on beaches across the world has increased. This actually means that plastic waste materials that were previously in oceans and seas are now surfacing and getting deposited along various beaches.

As mentioned above, several studies have surveyed beaches and water basins to assess the accumulation of debris along the coastline. In many of these studies, plastic products are identified as the leading debris item accumulating in the marine environment. The most common items that are found along the coastlines include plastic film, packaging products, straws, balloons, and metal beverage cans. More than 85% percent of the marine waste materials collected are usually plastics.

With these persistent qualities, plastic can continuously impact marine organisms through entanglement or ingestion. This has been publically documented and is a prominent concern. Since plastic can be found at a range of fragments, organisms from all levels of the trophic system can be affected. One way to address this issue is through examination of the impacts of human behaviour on this phenomenon. Very few studies in the social science literature have focused on the behavioural aspect of marine debris.

According to a statement by the Ocean Conservancy National Marine Debris Monitoring Program, "Only through changes in human behaviour and informed choices in products and packaging, can marine plastic pollution issue be effectively addressed and dealt with worldwide". For this reason, it is important to explore the attitudes and behaviours of people using marine waters in order to understand the attitude-behaviour relationship as it relates to the presence of plastic debris.

There has been a rich body of literature on environmentalism and how behaviours or behavioural intentions impact the environment to protect its natural resources. Behaviours aimed at reducing an individual's impact on nature are presented in many forms, but essentially all are cantered on protecting the environment. Sometimes terminology "environmental significant behaviour" is used interchangeably to mean actions which cause changes directly or indirectly to the environment. Pro-environmental behaviour is the "behaviour that consciously seeks to minimize the negative impact of one's actions on the natural and built world". It is also motivated by self-interest (such as doing what is right for the good of the population). These actions range from everyday performance to site- specific best practices.

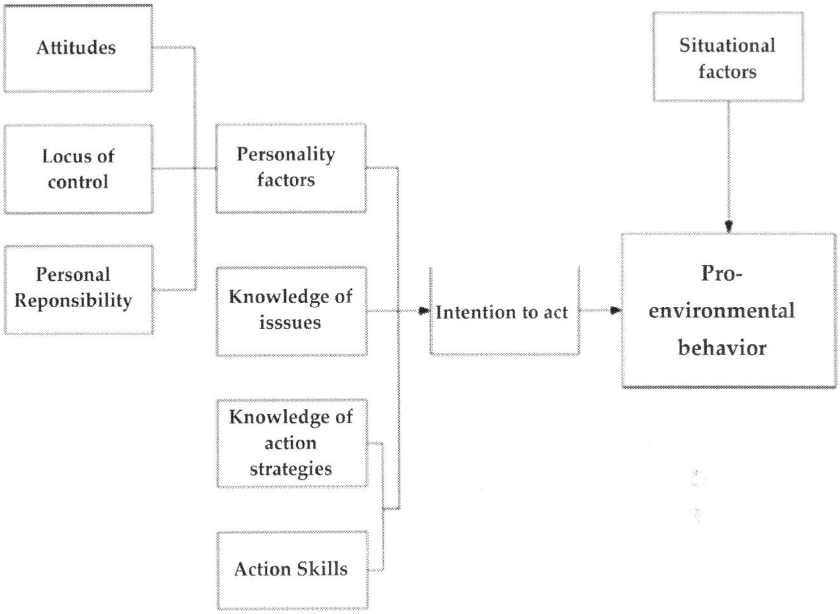

Source:https://www.researchgate.net/figure/235363126_fig2_FIG-3-Models-of-predictors-of-environmental-behavior-Hines-et-al-1986

Figure 4.12: Environmental behaviour model.

Traditionally in environmental education studies, it is assumed that if people become more knowledgeable on environmental issues they will develop more awareness of the problem and therefore become more motivated to behave more responsibly towards the environment. It is believed that environmental knowledge is an important component. This means that knowledge and awareness on the environmental issues are directly correlated to attitudes.

It is also important to incorporate the issue of recreation participation in changing the behaviour aspects of marine users. This is because recreational activities that people choose are usually consistent with how one values the environment and their attitudes of natural resources.

According to studies, it has been shown that recreation involvement usually acts as both a mediator between environmental attitudes and behaviours as well as a moderator within behaviour models. This means in a structural model; recreation participation is a variable having a direct impact on behaviour (moderator) or a person's attitudes are antecedent to recreation involvement which then has an effect on behaviour (mediator). Therefore, the inclusion of recreation participation as a predictor of both attitudes and behaviours has led to different results.

Lastly, much of the recreation literature also involves tourism studies since a person's motivation to visit a destination is based on leisure. Studies interested in tourist behaviour often explore recreation experience or involvement as direct correlates to that behaviour.

5 CHALLENGES OF PLASTIC WASTE MATERIALS ON DRY LAND

INTRODUCTION

Globally, plastic packaging materials generated are usually disposed inappropriately. There are several aspects that have led to worsening of this problem such as;

- Poor plastic waste management systems
- Single-use disposal nature of the product and its low cost
- Failure of institution and individuals to collect and dispose plastic wastes properly
- Absence of recycling activity and low public awareness on its management

Generally, plastic waste in most places around the world is acute problem, where waste management practice is very poor and it needs immediate solution to combat the threats posed to the environment. Thus, rising of public awareness on sustainable use and management of plastic waste is essential. Therefore, it is important to emphasize and promote the use of biodegradable plastic alternative packaging and carrier materials. It is also essential to promote reusing and recycling of plastic packaging materials in order to minimize plastic waste on dry land. In addition, the government should put a tax or ban on all single-use disposable plastic products.

According to UNEP (2005) report, it is important to note that the reason why most plastic materials have become dominant in most of the solid wastes in most places is because they are preferred by many retailers and consumers because of different reasons such as affordability, durability, light weight, and they are considered as the most hygienic way of packaging and transporting various goods and food products.

However, there is inadequate information on how many plastic bags have been produced and utilized in different countries across the world. This means that, the status of plastic waste disposal, its underlying causes and associated undesirable environmental impacts is not well known in most countries. Therefore, the primary focus of this chapter is to assess the plastic waste generation and its environmental impact on the land surface.

Knowledge on solid waste generation rate is important because it acts as a tool for proper management of wastes. Most of the plastic waste packaging materials are usually generated from residential, commercial, institutions and industries. According to UNEP (2005) report, the largest volume of plastic waste is generated from residential areas that constitute 55% of the total solid waste. The remaining 45% is generated from commercial, industrial centre and institutions.

Plastic waste generated from different places is usually disposed in open places. The disposal site is not designed to protect hazardous plastic wastes, thus all plastic wastes are usually indiscriminately disposed. The solid wastes disposed on the site are not covered with the soil. Therefore, plastic wastes particularly plastic bag waste escape from the site by wind and affect the surrounding environment.

Source:https://en.wikipedia.org/wiki/Landfill

Figure 5.1: A typical situation of solid waste materials in most places around the world.

PLASTIC WASTE GENERATION

Plastic waste quantities should be expressed in terms of weight and volume. This is because the latter varies with compaction. Plastic waste generation varies with time seasonally, monthly, weekly and daily. For instance, it has been proven that plastic and other solid wastes generation is usually higher during holiday seasons. The production of solid plastic waste also varies with population density. Urban life contributes the largest plastic and other solid waste generation, while rural people either makes with less and discards proportionately smaller quantities, or reuse much of the plastic waste. The reason for measuring solid waste generation rate is to obtain data to determine the total amount of wastes to be managed.

Uncollected plastic wastes are illegally dumped on open spaces, and water bodies. It is not uncommon to see littered food and other solid wastes into roadways, vacant place, and unpaved streets and disposed into rivers and drainage channels in most places around the world. This is evidence of mismanagement of responsible authorities.

Notwithstanding the fact that nature is capable of diluting, dispersing, degrading or absorbing plastic wastes, there might be stress on environment and hence an ecological imbalance will be imposed on the biosphere. For instance, inappropriate plastic waste disposal on the land serve as potential reservoirs of diseases. On the other hand, aesthetic considerations are related to the production of odours and the repulsive conditions that can develop when adequate attention is not given to the maintenance of sanitary conditions. Most odours can be controlled through proper maintenance of a reasonable collection frequency of plastic waste materials. It is also worth noting that most plastic containers that are improperly disposed usually act as key habitats of various diseases causing vectors such as mosquitoes and tsetse flies.

To calculate the overall average per capita generation rate, the waste generation of different income groups should be taken into account, since there is difference in mean household generation rate of plastic waste materials. The greater precision of weight average mean should be estimated by minimizing the sample error. Hence, the average per capita generation rate should include the percentage contribution in order to minimize errors. Therefore, in most places, especially in developing countries, the plastic per capita generation rate is usually more than 0.35gm/capita/day compared to the WHO standard rate of 0.223kg/capita/day.

CAUSES FOR PLASTIC WASTE

Durability of Plastic Materials

The majority of people in different countries usually do not have any permanent packing and carrier bags. Even those who have permanent bags are unable to use these bags for shopping. They prefer to use plastic bags rather than permanent shopping bags. This is due to plastic bags are convenient for handling and carrying goods. Moreover, they are accessed at any place and time in the supermarkets (shops). In addition, they are cheap and even given "freely" from supermarkets while buying goods.

Therefore, large numbers of people depend on using plastic packaging bags and making less effort to minimize plastic waste at the point of purchase. Most people use thin plastic bags more regularly than thick plastic bags. Thin plastics are a single-use disposable product and may not be used for a long time. Therefore, once used they are disposed in open environment and scattered with the help of wind. As the result, they have resulted to littered road sides, stream channels and open spaces.

To solve this problem, more environmental friendly alternative shopping bags (such as cotton bags) can be used to replace thin plastic bags or the community should have to use more durable thick plastic bags with reusing and recycling practice. These help to reduce the total amount of plastic consumed and lighten the environmental impact of disposed plastic materials.

The Cost of Plastic Materials

As shown in figure below, the costs of plastic bag determine the demand of the consumers for the plastic materials. That is "as the cost increases the demand for the plastic bag decreases."

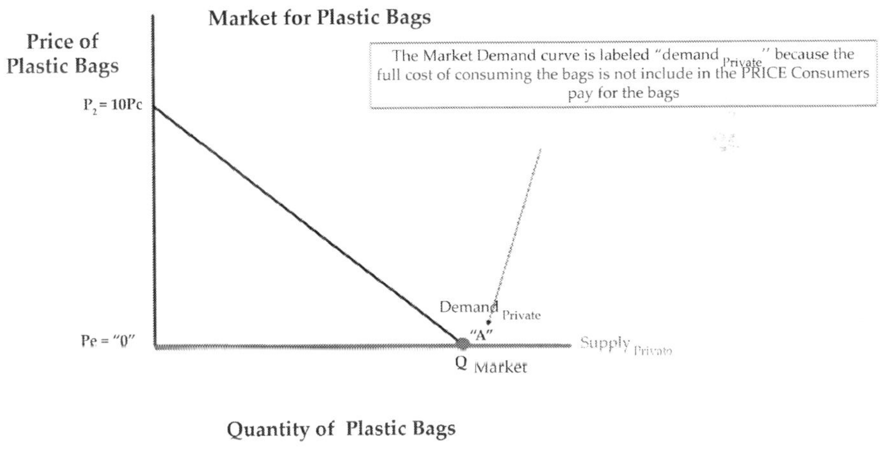

Source:http://haywardeconblog.blogspot.co.ke/2016/06/plastic-bag-tax-in-uk-correcting-for.html

Figure 5.2: Plastic bag demands with different costs.

Most people use thin plastic bags (bread bags) which costs a few cents or given freely by supermarket (shops). This indicates that, the price of plastic bag is an essential factor which causes plastic bag litters by leading to be used single disposable plastic products (thin plastic bag). The price that the consumer faces for plastic bags will impact on their use and re-use, the use of alternatives and the value people attributes to the bag after its initial use. For instance, in Denmark a levy applied to the consumer resulted to a 66% reduction in consumption of plastic packaging and carrier bags. This shows that putting a tax on consumption of plastic products is one of the options in order to reduce the impact of plastic waste on the environment. Therefore, consumer can reduce the number of plastic products that they used in a more effective and efficient way.

Moreover, proclamation on the use of plastic bag by Federal Environmental Protection Authority that prohibited granting of the manufacture or import of any non-biodegradable plastic bag with a thickness of less than 0.03 millimeters is an initiative that other governments can employee in reducing plastic waste challenges. All these regulations and proclamations have supported statements for appropriate control of plastic products. Thus, with the support of the concerned body, plastic waste pollution control measures are practically possible generally in any country.

Improper Management Systems for the Plastic Material

The final stage in the life cycle of plastic is disposal. In most countries, there are three common ways of plastic waste disposal (for instance plastic bags); by dumping them in the landfill, by burning them in homestead and by littering them. Even after reaching to the landfills, plastic bags escape from the containers and landfills by wind and become littered.

Open burning of plastic waste is a common practice in many households. This is the most inefficient ways of waste management with heavy environmental and health impacts. Besides open burning of plastics, most households use plastic for ignition of firewood. Generally, the failure of institutions and individuals to collect and dispose plastic wastes properly and absence of recycling activity is one of the root causes for occurrence of plastic litter.

Poor storage of plastic and other solid waste materials is another key challenge that usually escalates the issue of plastic littering on the land. The main equipment used by most households in the storage of plastic wastes and other solid materials before disposal is paper bags and plastic bins. Both of which are provided by private contractors and Community Based Organizations (CBOs). Other means of storage include cartons, baskets and synthetic sacks.

Use of paper bags for storage is an expensive method. Use of bins on the other hand is an alternative that reduces costs of storage both for households and the collecting firm. This is because bins are much cheaper for waste storage than paper bags.

Transportation equipment is another challenging issue with regards to management of plastic waste materials. Transporting of waste in most countries happens in two steps. Firstly, after plastic and other solid waste materials are collected from households, they are transported to collection points. For instance, in a country such as Kenya, plastic material scavengers are mainly responsible for this part of the process. Secondly, from the collection points, plastic waste materials are gathered and taken to the final disposal sites. This is done by the private contractors who are contracted by the municipalities. The role of the municipalities includes collection, transportation and proper disposal of plastic and solid wastes; however, in some countries such as Uganda and Ethiopia, the municipalities are not directly involved in collection and are barely involved in transportation. They act by contracting private companies who collect waste from collection points.

The private contractors collect waste from the collection points to the final dumping grounds. In developing countries such as Kenya, there are inadequate private contractors who have ventured into this business. This leads to waste accumulation in collection points. This phenomenon is quite common in most cities and towns.

Another challenge that is usually faced is inadequate personnel. Solid waste management is an intensive field of service that requires a large number of human resources for the service to be rendered adequately. Studies have shown that human resources needed for plastic waste collection are not sufficient in most places. It should be noted that county governments in most countries are usually not directly involved in waste collection. Despite the fact that most county governments usually have inadequate human resources, it is their responsibility to work together with other municipalities to improve plastic solid waste collection and proper disposal. Private contractors do not have a major problem with personnel since they only collect waste from collection points to the disposal sites. They only require drivers.

Public Awareness

Plastic packing materials and other plastic materials are the common litters in the waste stream. Most people know that landscapes littered with plastic bags are hazardous to domestic animals and visually unattractive. Nevertheless, are usually unaware of the effect of plastic bags. In some pastoralist communities, domestic animals have a probability to take plastics while grazing on the land.

Generally, most people know some of the impacts of plastic to the environment, there is the need for additional public awareness creation in order to bring behavioral change in the community for managing and controlling of plastic bag pollution.

MANAGEMENT OF PLASTIC WASTE MATERIALS

Plastic waste materials can be defined as waste coming from plastic packaging materials, plastic food waste containers, plastic packaging bags, synthetic textiles and even plastic ash waste. Plastic waste materials can be generated from cleaning process in the streets including drain cleaning and plastic litter that is collected from the main land. Most plastic waste materials are believed to be coming from residential areas, offices, hotels, schools and even social complexes.

The rise in population over many cities and towns and industrialization has become an increasing contributor to plastic waste discharge. Until recently waste management was never held in high regard. It was a subject that was not considered as important in many municipalities and government offices especially in developed countries. Increasing number of industries producing different kinds of plastic packaging has been a major contributor contributor to environmental pollution, which has led most governments to control the plastic waste materials discharged into the natural environment.

Environmental issues have attracted the attention of many people. People are becoming increasingly conscious of a variety of problems like global warming, acid rain, pollution in air, water and land etc. All these issues have been of concern to many nations and citizens at large. Plastic solid wastes result from human domestic activities, social and industrial activities for the purpose of improving the standard of living increase in population and development of technology.

A major problem encountered in big cities is plastic waste management. Plastic waste management is the most important among all the other wastes such as municipal, industrial and healthcare wastes. There is often lack of proper planning and management thus leading to immense pollution of water, soil and even atmosphere in the surrounding environment.

Source:http://blog.nus.edu.sg/plasticworld/2016/09/21/9-plastics-and-air-pollution/

Figure 5.3: Illustrating air pollution through incineration of plastic waste materials.

A great problem that the world is facing today is the cost of protecting the environment in relationship to the return that would be yielded from that cost. The current practices of controlling plastic pollution, plastic waste treatment and proper plastics disposal or recycling mechanisms are considered very expensive ventures. Traditional approaches to plastic wastes management could be actually considered disastrous since they may result to depletion of natural resources due to the fact that the disposal is not done properly. This is the reason why proper sustainable plastic waste management measures should be taken to curb the issue. Plastic waste generated in very large quantities is difficult and expensive to dispose. Most of the large sources of such wastes are industries and residential places. Such plastic waste will always pollute and intoxicate the environmental resources in a way that causes a threat to the health of human being and even animals and plants. The dynamic and ever changing environment leads to many new and unique types of wastes produced; the challenge becomes the cost of managing and disposing of this waste in an environmental friendly manner.

To attain sustainability in management of plastic wastes, there are certain elements that must be considered. These include;

- Social elements
- Conservation perspectives
- Change of behaviour

These are the main areas that form the basis of environmental sustainability. According to UNEP (2005) report, for any solution related to environmental sustainability to be successful there should be focus towards economic benefit, available technology and it should comply with social and environmental norms as defined by each specific country.

It is believed that sustainable management solutions would employ regulations that would reduce plastic waste at the source, reuse, recycle, recover and ultimately rethinking and renovating whereby people should rethink about the plastic wastes they generate closely before taking any actions to eliminate them. They should hence innovate by coming up with innovative measures of disposing or recycling the plastic waste in an environmentally friendly manner and possible derive value from the waste.

There are numerous approaches that can be employed in plastic waste management. For instance, a top down approach can be employed in plastic waste management. For this approach to be effective, there are certain elements that should be incorporated into the approach such as ensuring that there is adequate engagements with the citizens, community ownership, and users involvement. Community involvement is necessary in order to reduce conflicts and resistance, especially when implementing plastic waste management strategies.

Studies have shown that capacity building, focus group initiatives and access to information are important elements towards plastic wastes management sustainability. In addition, it is also important to carry out awareness campaigns and allow citizens to participate and also provide solutions to waste management. The deficiency mentioned by previous researches includes the lack of a comprehensive system of managing plastic waste. Most studies proposed a single solution to achieve a final disposal of all kind of plastic waste (both industrial and residential). It has hence been proven that an integrated system of managing plastic waste is critical in ensuring sustainability in plastic waste management.

It is important for responsible governments and other non-governmental agencies to put up sustainable waste management systems that will be practical and environmentally, economically and socially acceptable. An integrated approach to plastic waste management is paramount. Implementation of an integrated waste management decision support system is significant. This model should be all round in order to cater for social, economic and environmental factors. This should include source separation methods, plastic waste collection, treatment, disposal and any potential environmental impact that may be experienced.

Over time, plastic waste management has been a forgotten topic in most organizations. Top management often focuses on increasing revenues, reducing costs and hence improving the profitability of a business. However looking at the social, environmental and legal implications, plastic waste management has recently been a problematic area that has threatened to have serious financial implication from an environmental and legal perspective. Hence, proper management to ensure sustainability is of great importance.

Challenges in Management of Plastic Wastes

Most developing countries have developed different Solid Waste Management (SWM) projects with the help of national government and other non-governmental organizations. Many of these projects are often funded; hence once the funding is completely depleted, these projects fail since they are not able to self-fund themselves hence there is a risk of un-sustainability of such projects. For instance, in Africa most of plastic wastes are deposited at dumpsites and most of the adjacent areas are quite disadvantaged due to the many operations that are undertaken at the dumpsite. The existence of these dumpsites in cities and major towns has led to several issues like poor infrastructure, including roads and drainages, and unending dumping of plastic wastes. In addition, most dumpsites have posed significant pollution of adjacent resources (such as rivers), air and soil pollution is also exhibited. Cities dumpsites have also posed risk on the health of people and domestic animals living around there. Plastic waste separation at source is a significant area of focus that should be recommended in ensuring sustainability of plastic waste management. In most countries, plastics waste can be of great value if separated at source and availed to focus women groups who creatively use the waste to make nice jewellery, hats, baskets, door rugs and mats. These items can be sold as unique crafts to the local community and hence act as income generating source for such women groups.

Source:https://www.changemakers.com/economicopportunity/entries/garbage-baskets-income-generation-re-generation-next-ge

Figure 5.4: Women undertaking basket weaving using plastic waste materials.

In essence, the problem of plastic waste has continued to suffice in the modern environment. This has therefore translated to high wastes associated costs as charged by governments on the production of plastic wastes through various industrial processes. Governments therefore need to adopt and implement plastic waste management practices owing to the fact that plastic wastes will always be there. Nevertheless, optimal productions processes will significantly reduce plastic products at source thus improving on the chances of re-use and re-cycle of plastic wastes materials (such as plastic water and beverage bottles). Production will be able to continue undeterred and at the most optimal efficiency level that strives to cut plastic wastes significantly.

PLASTIC WASTE REDUCTION STRATEGIES

Reduction of Plastic Waste Generation at Source

One of the sustainable methods of ensuring sustainable plastic waste management is the integrated method which seeks to reduce waste at source before it goes on to the next level. It also involves looking for opportunities to reduce plastic waste even during the product life cycle. Once plastic waste is generated, then it is recovered for reuse and if there is still waste, then it is finally recycled. In addition, methods are put in place to look for more environmentally friendly materials. Plastic waste reduction or minimization is critical in effectively managing plastic waste. It is most important to reduce the waste from source.

There is the need for implementation of simplified leanness evaluation metrics to reduce plastic wastes and improve effectiveness and efficiency in a place. Continuous Performance Measurement is an important matrix that should be used to assess leanness reduction of plastic waste generation.

Creating Plastic Waste Awareness

Successful plastic waste management is often attributed to many reasons arising from policy implementation; however the main reason why most countries have succeeded in management of plastic wastes is due to creation of public awareness and support. According to World Health Organization (2010) suggestion, involving communities in plastic waste management programs may help in promoting publicity with tips on waste management hence eventually minimizing plastic wastes.

Similarly, UNEP (2010) report indicated that two main underpinning group of drivers of plastic waste management include public awareness and responsibility issues. Awareness creation is critical in ensuring that there is sustainability in plastic waste management. It starts from people knowing which plastic waste elements exist in a place and which methods are to be used to effectively manage, reduce and properly dispose the waste. One of the challenges facing proper plastic waste management, especially in developing countries is the lack of proper public awareness on plastic waste reduction. This is also observed by the fact that the general public usually have negative attitudes towards plastic waste management; hence the government should carry out campaigns to enlighten the public on plastic waste and plastic waste management.

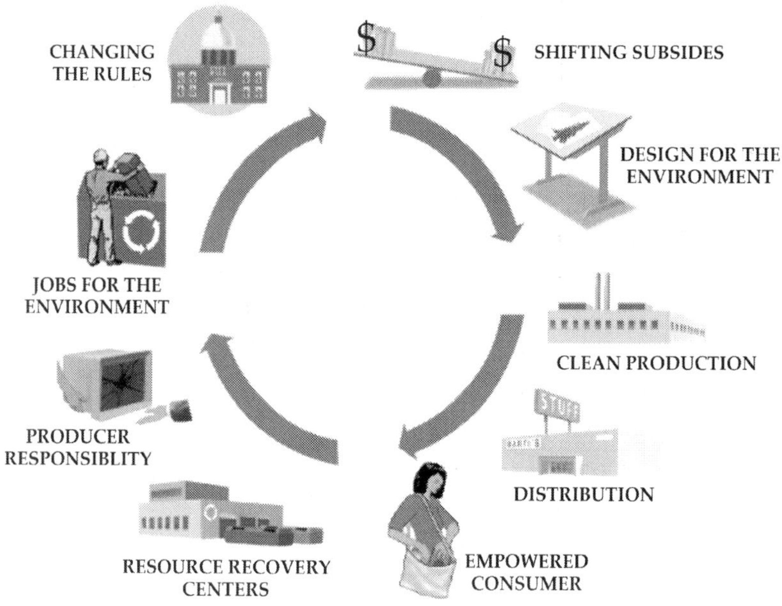

Source:http://www.improntaunika.it/2014/08/conca-park-hotel-albergo-rifiuti-zero/

Figure 5.5: A typical zero plastic waste system cycle.

Studies have also shown that education is a significant element that can be used to encourage the general public in participating in plastic recycling programs. It is therefore important to advocate for plastic waste educational programs that would ensure that communities are involved by increasing awareness and commitment towards plastic waste, increasing capability of different staffs in identifying opportunities that would lead tominimization and avoidance of plastic usage and ensuring that operational staff are adequately trained for foster compliance with relevant plastic waste regulations and are able to report any negative implications or observations.

Awareness of impacts of plastic waste has not been enforced in most countries. For instance, studies have revealed that many people are usually ignorance and are involved in littering everywhere with plastic waste materials. This is evidence of lack of proper awareness in most areas when it comes to plastic and other solid waste management. According to Department Environmental Affairs (2011) report, enforcing anti-plastic littering campaigns and plastic waste separation at source is essential to plastic waste management sustainability.

Plastic waste awareness and participation can also be enhanced more by creating recognition programs such as the cleanest town competitions, coming up with specific performance evaluation criteria and reward and recognition. Additionally, plastic producers and consumers responsibilities must be aligned to the overall waste management plans and consumer awareness programs should also be incorporated in plastic waste management plans. Plastic waste awareness can also be created through door to door awareness and motivation programs which involve establishing contact with participants and providing feedback, it also helps in reducing the time lag that would be created between information communication and when the actual waste collection begins. Importance of door to door awareness campaigns is that communication is effectively passed on to the participant and often reduces any elements of rumour mongering, building confidence of the participants and also assisting in clarifications of any issues of concern.

Plastic waste awareness can also be created through rallies as people get very excited with rallies and street games and they tend to attract a large crowd to come and witness what it is all about. Embedding plastic waste management into school programs is also a plus since children are the future of tomorrow, this is a more sustainable tool toward effective plastic waste management. Motivating individuals towards plastic waste intolerance is a plus since the individuals will exert pressure to the companies and authorities that will ensure proper management support and implementation. Public education and waste awareness is also a critical method of ensuring management and sustainability of plastic waste materials.

Waste Management Policies and Legislature

One method to ensure effective plastic waste management is having proper management systems which are included in the policies and procedures within the responsible authorities. For instance, the most effective ways of minimizing or reducing plastic waste is through introduction of a tax policy that will ensure that individuals are taxed according to the total amount of plastic waste they produce. Additionally, higher taxes can be imposed on raw materials to encourage manufacturers to produce goods that generate less plastic waste. Studies have indicated that reduction of VAT on plastic packaged items with environmental friendly labels could help in minimizing the total amount of plastic waste from both producers and consumer perspective. Goals for plastic waste prevention and incorporation of such goals in the waste strategies and policies are significant in promoting sustainability. It may be tough for the responsible governments to enforce plastic waste prevention measures when dealing with consumption but it is definitely a key measure when it comes to legislation and policy making.

Nevertheless, one of the key challenges in management of plastic waste material is the inability to actually make people to stop generating plastic waste materials. This is the initial step to effective plastic waste management. Instilling waste discipline is another loophole, especially where there is no system in place that would ensure people are responsible for the plastic waste they generate and that they are liable for their actions. Putting up effective systems to ensure individuals are held accountable for the amount of plastic waste generated would be a plus. Environmental assessment audits are also not adequately done in most countries. These assessments would be significant in identifying whether individuals and companies comply and that they adhere to the safety standards.

Reduction of Plastic Consumption relates to 'throw away' Society Topic

In addition to reducing plastic wastes based on the above strategies, it is important to also look at the broader topic of how the society can change the 'throw away' perception on plastic wastes and the need to alter this behavior. According to World Bank (2010) review, currently, many people live in an economy where 80 percent of what they consume gets thrown away after one use. This mentality is evident from the amount of plastic in the environment as depicted by the Great Pacific Garbage Patch (GPGP). The GPGP is a portion of the ocean that is plastic rich with marine debris, especially single-used plastic packaging bags, and the debris is captured in a product of swirling currents called the North Pacific Subtropical Gyre. The GPGP is found in the North Pacific Ocean because all the world ocean currents are connected and congregate in that area.

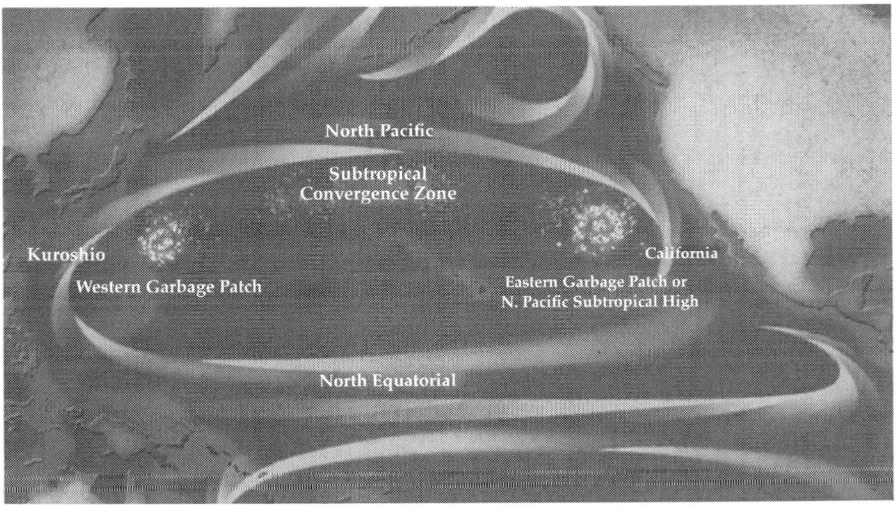

Source:https://marinedebris.noaa.gov/info/patch.html

Figure 5.6: Illustrating the Great Pacific Garbage Patch.

The GPGP was discovered in 1997. This soup of trash is 90 percent plastic and 10 percent free-floating fishnets. A study done in December 2014 by 5 Gyres Institute estimated 5.25 billion plastic particles weighing about 269,000 tons are floating on the surface of the sea. Of the plastic debris, single-use plastic packaging bags and plastic packaging bottles are main contenders found. According to UNEP (2010), about 80 percent of the plastic debris originates from land sources while the other 20 percent originate from sea faring vessels and oil platforms.

The GPGP is not only a huge problem for organisms in the environment but microplastics can also be transferred from organism to organism by ingestion and can move up the food chain, ultimately affecting human health. This topic has attracted interest and funding from the US Environmental Protection Agency (EPA), the National Oceanic and Atmospheric Administration (NOAA), and the National Academy of Sciences (NAS), as well as other researchers, nonprofit groups, institutions around the world. However, more research needs to be done to fully understand the effects of microplastics moving up the food chain, to determine how toxic it is towards human health.

These issues are as a result of a 'throw away' society and to combat this issue, policies must be made to reduce plastic production and consumption. Reducing plastics consumption is a good first step in raising awareness among locals and tourists about the harms plastics have on the environment and human health. Furthermore, reducing plastic usage will hopefully spark a conversation to challenge the production and consumption of other items such as Styrofoam cups and containers, plastic cups and bottles. These items also pose a major threat towards the health of natural environment and towards human health.

According United Nations (1987) suggestion, to successful reduce plastics consumption, there is the need to be a balance between three sectors to be sustainable meaning "development meets the needs of the present without compromising the ability of future generation to meet their own needs". To achieve sustainability, the following three sectors must be taken into account;

- Environmental sector
- Societal sector
- Economical sector

For instance, if a country reduces the use of plastic because of environmental and economic reasons without the support from society, no citizen will abide by the laws and then the policy will be a failure.

Plastic Waste Reusing and Recycling Strategies

Many countries that have incorporated plastic waste management in their sustainability plan often reap a lot of benefits. A country like United Kingdom has adopted a plastic waste management hierarchy that helps in reducing plastic waste obligations as much as possible. Reusing plastic wastes (such as plastic packaging containers) is the process of using materials repeatedly. In other word, this is where an individual or company takes a plastic waste material and after it has gone through its useful life it is used again for other purposes.

Plastic wastes are representation of both material and energy loss. When so much plastic waste is generated, this is a sign of inefficient processes in a country. Plastic waste is generated through several human activities. This is because over the years people have a desire to have a better quality life. Studies have indicated that reusing plastic waste occurs when the waste destined for disposal is reused. A good example is obtaining plastic soda bottles and using them to store milk or water. This drives sustainability.

Plastic waste reuse: Several countries have different kinds of plastic waste generated from their industrial and manufacturing processes. To minimize plastic waste as much as possible, this can be done from the source where the plastic materials are coming from. Plastic waste reduction is always a factor that enhances waste sustainability. According to WHO (2011), plastic waste should be avoided.

Another effective way to reuse plastic waste is through burning in incinerators. The resulting residue after such incineration can be dumped in landfills. Combustion can be defined as the burning of waste in a controlled environment in a bid to reduce waste volumes that are sent to landfills. This process can also generate electricity. This technique is effective especially where plastic wastes cannot be recycled and are generated in large quantities. Important to note is that the plastic waste combustion process can lead to the emission of toxic air emissions, but they can be controlled through the installation of control equipment for instance fabric filters. According to a research study that was conducted in the United States of America, it was revealed that about 15% of plastic wastes produced are not just used to burn trash, however the heat generated is used in generation of electricity.

Incineration reduces plastic waste volumes by 80% to 90%. The salvaged plastics together with generated waste energy as a result of combustion can be consequently used to generate readily affordable electricity and even heat to nearby buildings. Incineration does not pollute underground water reserves, it also adds very little to current levels of air pollution if proper equipment to control air pollution during combustion are installed as per prevailing environmental laws. Since urban areas are running out of enough landfills, incineration will be the most preferred choice to dispose plastic and other solid wastes since it is way much economically attractive than maintaining current limited landfills.

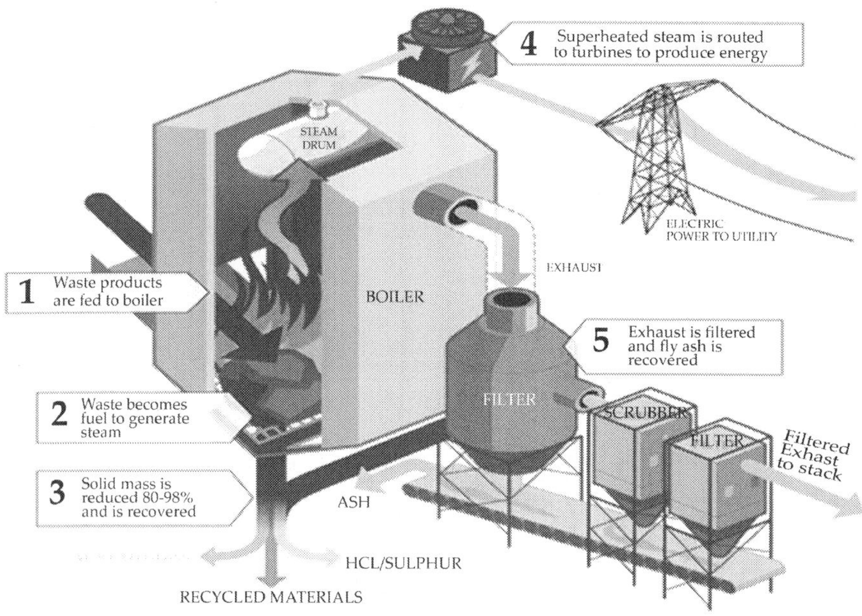

Source:http://www.ecolog.com/issues/verify.asp

Figure 5.7: Illustrating a typical power generation station using plastic waste combustion.

General public participation: Studies have revealed that, plastic waste management in most places, especially in Africa and other developing countries is carried out by cart pushers, resource merchants, private waste collectors, public, and neighbourhood and estate associations. It is also worth noting that most County Governments in different countries usually collect and then transports waste to a specific dumping site. Collections, transportation, sorting, financing and recycling are some of the major tasks that the general public usually undertake to manage plastic wastes in different places around the world. Potential roles that the general public play are waste separation, composting, distribution of plastic waste containers and subsequent re-usage of collected and separated wastes. Various challenges however suffice in attempts to enhance sustainable waste management.

- For instance:
- Inadequate resources
- Averseness
- Poor attitudes
- Plastic waste management knowledge gap

It is therefore important to adopt different methodologies to ensure sustainable plastic waste management; public sensitization, the enhancement of social interactions with like- minded individuals, prioritization of plastic waste management during countries' annual budgets, and subsequent plans to close the existing knowledge gaps of plastic waste management.

To ensure homesteads readily participate in plastic waste re-use, quality of environmental education, efficient plastic waste collection regime, attitudes and subsequent enforcement scheme ought to be enhanced. Plastic waste collection regimes therefore need to receive adequate environmental education and attention in order to ensure that the public embrace plastic waste re-use with ease. This will consequently translate to a ready market of re-used products thus ensuring that the demand levels for such related products are sustainable. If the public can be associated and readily involved in plastic waste re-use projects so as to contribute their own efforts, sustainability of plastic waste management programs can be achieved hence yield success. Relevant authorities ought to consider the social and economic status of the community involved in plastic waste management so as to ensure that started waste management projects are successful in the immediate surroundings.

Reuse in relations to environment: Plastic waste material recycling mostly depends on the policies that have been enforced on recycling. There are several advantages of recycling. For municipal, plastic waste material recycling helps in the overall reduction of the waste volume, there is a lot of cost saved from handling, collecting, transporting such waste and disposal of the waste in general. The economy will also benefit from recycling plastic wastes since more people will get employment. The environment is the overall beneficiary of recycling plastic waste since there will be an overall sustainability of environment and waste going into storage sites will be reduced resulting to a more manageable system.

Therefore, there is need to ensure flexibility in plastic waste management systems with regards to the ever dynamic economic, environmental and social conditions. Many a times there are several processes involved in plastic waste management which are closely related and hence present an integrated approach which is better to design other than having individual systems that are not value adding.

Plastic waste segregation at source is a key step towards effective recycling. Segregating plastics from paper at source can enable companies to find better ways of recycling paper and plastic. Recycling a collection of different solid waste is one of the most difficult activities in the recycling process. These could be mixture of paper, plastic and even food stuff. It is difficult to effectively recycle waste when separation or segregation is not properly done at source. In addition, mixing different types of wastes leads to poor quality of the products being recycled. Segregation of all types of waste at source is the first and most important step that will lead to saving of resources such as time and labour when it comes to plastic waste recycling.

Waste Technology: The use of technology is important, especially when it comes to sustainable plastic waste management. Improving technology, especially in developing countries would be of essence. This is because most developing countries lack proper plastic waste management systems leading to short term and long term impacts on employees and on the country at large. Governments ought to put up proper management systems that would bring solutions geared towards plastic wastes reduction, recycling and disposal as well as proper collection and transportation methods. Emphasis should also be placed on costing of such waste management systems and inclusion of waste costing in the countries' budget. Both producers and consumers ought to get involved in Corporate Social Responsibility (CSR) initiatives that address pollution and waste management.

PLASTIC WASTE MANAGEMENT IN THE INFORMAL SETTLEMENTS

Studies have shown that there is accelerating rural-urban migration as people search for jobs and other opportunities. This explains the reason why most cities' capacity is overwhelmed in providing standard services and resources to their inhabitants resulting to development of informal settlements. The lingering questions to ask are;

Is the government anticipating the population growth?

Is the responsible government working towards realising equal distribution of resources and services in order to meet the needs of all citizens the same way taxes are collected from all?

Are there efforts to decentralize services and resources held in the big cities to minimize the rural-urban or/and urban- urban migration?

Plastic waste management in informal settlements basically involves plastics waste containers, plastic packaging bags, and other types of plastic wastes generated in informal households. Plastic wastes can further be divided into municipal and non-municipal plastic wastes.

Municipal Plastic Wastes

According to UNEP (2005) report, municipal plastic (solid) waste can be defined as all household plastic wastes, business plastic wastes of small amount less than 50kgs, various plastic wastes generated from public institutions such as schools, market places and hospitals that do not require any treatment, and street plastic wastes.

Municipal services are lacking in most of the informal settlements, especially in developing countries which are marked with huge dumpsites and blocked drainage systems. According to UNEP (2005) report, as population grows, material consumption increases resulting to more wastes generated.

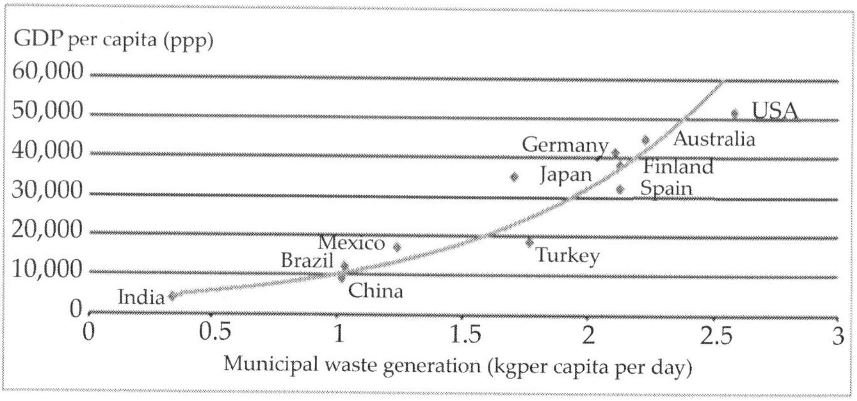

Source:http://www.businessinsider.com/waste-and-gdp-per-capita-around-the-world-2015-6?IR=T

Figure 5.8: Illustrating municipal plastic wastes and GDP per capita.

Non-municipal Plastic Wastes

Most of the non-municipal plastic wastes generated in the informal settlements are not different from municipal plastic wastes. Some of these wastes include plastic bags, food packaging plastic containers, synthetic clothes materials and other household plastic waste materials.

The major source of income in most of the informal settlements around the world is informal jobs in the small business enterprises. This has resulted to a uniform pattern in plastics and other solid wastes generation at the household level. Research findings have revealed that income levels of households highly relates to the types of wastes being generated. For instance, higher income households usually generate more domestic solid wastes like plastic bags, plastics containers, and synthetic waste cloths generation as opposed to those with lower income levels. According to World Health Organization guideline values, it is suggested that each person is supposed to produce 0.5-1 litre of refuse per day with an organic content of about 25-35% and a moisture content of 10-60%. This however is different from informal settlements where individuals generate about 1.7-3 litres in their households. In addition, studies have shown that households with higher female members usually have an upsurge on the types of plastic and other solid wastes being generated. According to UNEP (2005) report, it was concluded that generation of the plastic wastes has increased so much due to the current trend whereby most groceries and other commodities are packed in plastic polythene and/ or tins. Most governments in conjunction with other non-governmental organizations have taken efforts to ban use of flimsy plastic bags with the aim of minimizing the over increasing challenge in management of plastic wastes around the world.

Most of the people in informal settlements (80%) usually store their plastic and other solid waste in shallow rubbish pits outside their houses. With time these pits turn into large dumpsites due to irregular waste collection services. Studies have indicated that many of these pits are usually not fenced off. Most of them are drainage channels turned into rubbish pits resulting to blocked drainage systems. There is also the use of plastic dustbins in some of the households; most of them are usually old broken buckets or pans recycled as dustbins.

Source: http://sulabhenvis.nic.in/LatestNewsArchieve.aspx?Id=5228&Year=2013

Figure 5.9: Illustrating drainage channels turned into rubbish pits.

In most countries, the municipal councils usually place plastic bins in some strategic places but most of these bins are usually not managed properly due to large amounts of plastic and other solid wastes from households in the informal settlements. However, the use of polythene papers in plastic wastes storage is quickly overtaking the use of bins. Some of the households reuse the polythene papers they get after shopping in collecting wastes. These are later disposed in the nearby dumpsite.

According to WHO (2005) report, there are several characteristics of a good storage container that should be followed. These characteristics should include;

Storage containers for municipal plastic and other solid wastes should be made of durable, watertight, rust-resistant material

Should have a closely-fitting lid and handles to facilitate collection

The size of the storage containers for residences and businesses should not be less than ten gallon nor more than thirty-two gallon capacity, unless the containers are of the type that can be mechanically lifted and emptied by a collection vehicle, in which case they should be of appropriate size and design

- All storage containers should be leak-proof
- The interior of a container should be kept clean by thorough cleaning and sanitizing as often as necessary
- Storage containers should be kept in good repair
- Containers with loosely fitting lids or doors or containing holes or cracks should not be used
- All storage containers designed for mechanical lifting should be designed, constructed and placed to prevent accidental overturning
- Storage containers should be constructed of fire retardant material

In summary, it is clear that majority of the residents in informal settlements mainly use dust bins or rubbish pit to store their plastic waste and other solid wastes. On the other hand, some households also use polythene bags in storing their plastic and other solid wastes. According to UNEP (2010) report, some of the improved storage facilities used in some informal settlement across the world include large municipal bins, household plastic containers, and other communal rubbish bins. However, majority of people in the informal settlements usually turn into open dumping sites and this has resulted to ever increasing problems of scattered plastic wastes materials on dry land.

6 KEY CHALLENGES OF PACKAGING PLASTICS ON HUMAN HEALTH

INTRODUCTION

In recent decades, plastic has become the magic bullet of industrial production, highly valued for its durability, cost-effective synthesis, and utility as a packaging material. Ever since the development of the first fully-synthetic plastic by Leo Baekeland in 1907 and its subsequent chemical refinement, plastic has come to fulfil myriad industrial, technological, textile, and packaging functions, used in everything from food containers and automobiles to medical procedures and children's toys. As a synthetic organic polymer derived from petrochemicals, plastic is a particularly resilient building material and has come to displace many traditional materials such as wood, stone, glass, leather, metal, and ceramic due to its malleability, ease of manufacture, and resistance to water. This strength and impermeability also renders it a particularly valuable material in food and drinks packaging containers, shopping bags among other functions. Furthermore, plastic mixes well with other chemical derivatives, resulting in an ever-expanding diversity of industrial and production uses.

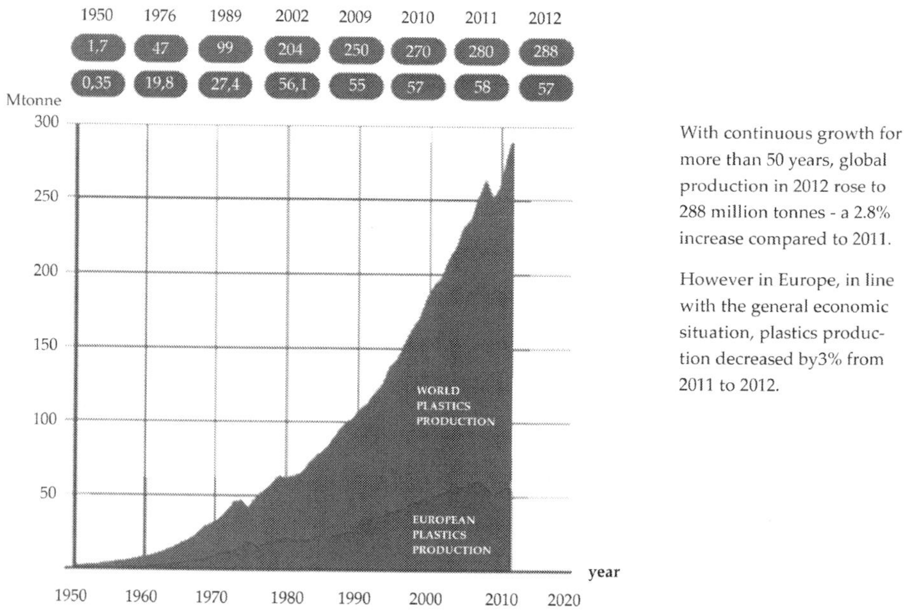

Source:https://discardstudies.com/2015/10/03/the-ocean-conservatorys-call-for-mass-incineration-in-asia-disposability-for-profit-fantasies-of-containment-colonialism/plastic-production-world/

Figure 6.1: Illustrating world's plastic production.

This proliferation of plastic production along with a greater understanding of its chemical impermeability has given rise to increasing environmental concerns regarding its slow decomposition and its tendency to be discarded in landfills in large quantities. Indeed, the birth of the recycling movement was largely in response to scientific and public recognition of plastic as an environmental contaminant, facilitating a public attempt to reduce the amount of plastic waste being generated.

Nonetheless, recent plastic optimization techniques and higher product demand have rapidly accelerated global plastic production to nearly 300 million metric tons annually. Of the plastic that is discarded, only about 6.5% is recycled into new materials and only 7.7% is used to generate electrical and heat energy; an estimated 30 million metric tons are projected to end up in landfills or leak directly into water systems. According to environmental scientists at Columbia University, low recycling rates are likely due to inefficiencies within the recycling process (exacerbated by plastic waste contamination and confusing coding designations by the Society of Plastic Industries) and public ignorance of correct recycling practices. Indeed, plastic debris is now so ubiquitous it is thought to contaminate almost every ecosystem on earth.

Plastic wastes account for 12.33%, 12.79% and 13.64% of the total amount of solid waste produced in low, middle and high-income residential zones respectively. Nearly half of these are composed of flimsy as well as thicker plastic bags. Plastic products are non-biodegradable and once they are disposed, they take an indefinitely long period to break down. Incineration or burning of plastic wastes emits poisonous gases and fumes into the atmosphere leading to air pollution. This causes respiratory health problems to human beings.

The issues of air pollution due to plastic waste combustion in dumpsites have been talked about for a long period of time. A good number of authors have written about this topic and even further on the topic of plastic waste management. There is therefore a substantial amount of literature on this topic. It is very clear that the amount of plastic waste produced will increase as time goes by due to population explosion and urbanization. Although there has been some success in protecting the quality of the environment in general, the challenge lies in waste. There is a need to advocate for more environmentally sustainable methods of plastic waste disposal.

Plastic waste disposal is a smaller topic in the grander topic of solid waste management. It refers to what is done to the plastic waste to get rid of it.In other words, it involves a number of processes like plastic waste collection, recycling and transport among others. Many governments in most countries worldwide have seen the need to legislate the process of plastic and other solid waste management through statutory provisions and by law. This is a bid to control and regulate the various processes involved.

On the other hand, the government in its endeavour to bring under control the problem of solid waste disposal should seek measures to cut down on the amount of plastic waste produced countrywide. Emphasis should be placed much more on the control of production of waste that is not environmentally sustainable such as plastic packaging containers and bags.

According to a recent UNEP (2010) report, it was revealed that more than 100 million plastic packaging bags are supplied by supermarkets alone. This is just a fraction of the total amount of total plastic packaging bags and other containers. According to World Health Organization (2006), the situation in European countries as far as plastic waste is concerned is as follows:

Total plastic consumption including virgin polymers and recycled granules continued to increase, rising from 6% between 2000 and 2002. The significant growth occurred during 2002 (4.1%) and only 2% during 2000 to 2001.

The per capita consumption of virgin plastics in the Western Europe rose from 91.5kg to 94.8kg during the period 2000 and 2002.

In 2002 the total collectable waste recovered was 38%, it was 36% in 2000 and in tonnage terms, it remained the same at 11%.

Mechanical recycling of plastic waste was 17% in 2000 and it increased to 22.5% during 2000 and 2002 in tonnage terms. There was significant increase in recycling of packaging waste.

The energy was recovered from 4,411,000 tons of plastic waste in 2000. It increased to 4,688,000 tons in 2002. An increase of 6.8% was recorded. However, it is not encouraging to note that the plastic waste going to landfills also increased a little bit.

DEFINITIONS AND CHARACTERIZATIONS OF MICO-PLASTIC PRODUCTS IN THE ENVIRONMENT

According to World Health Organization (2006) report, micro-plastics are small plastic particles <5 mm in diameter; however, further characterization has diversified them in terms of anthropogenic origin, shape, size, density, and type, making them particularly difficult to study. Microplastics may be derived from a variety of industrial and postconsumer sources, including but not limited to exfoliating cosmetic products, chemical run-off from processing facilities, and most significantly, in-situ degradation of larger plastic pieces deposited along shorelines and at sea. Those particles which exist and are deposited in their original microplastic form (e.g., scrubbing beads in facial cleansers and pre-industrial "nurdles" used as thermoplastic processing feedstock) are given the designation "primary microplastics" while those that are as a result of photodegradation and weathering of various large plastic pieces are described as "secondary microplastics". Additionally, microplastics may assume a variety of physical shapes (such as fragmented shards or shavings, filaments, pellets, and spherical beads) and sizes.

Most of the plastic materials usually exist in a number of types, including high-versatile polymers that are used for packaging and preindustrial building materials (such as polyvinyl chloride (PVC), polypropylene (PP), and polyethylene (PE), which represent 19%, 24%, 38% of the total global production respectively).

Polyethylene in particular is a critical and increasing source of microplastic contamination; with over 80 million tons of polyethylene generated each year, it is by far the most common plastic used in industrial and packaging production. In its unreactive form, polyethylene is a white waxy plastic composed of a series of nonpolar saturated hydrocarbon chains, rendering it particularly stable, hydrophobic, and resistant to chemical and physical degradation.

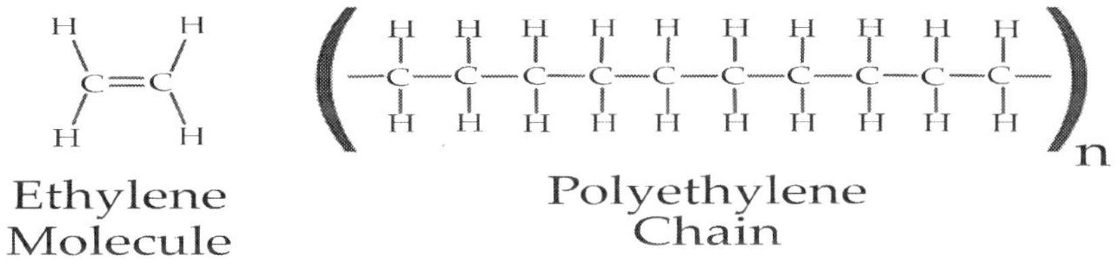

Source: https://www.learner.org/courses/chemistry/text/text.html?dis=U&num =Ym5WdElURS9OQ289&sec=YzJWaklUQS9OeW89

Figure 6.2: Illustrating a polyethylene chain formation.

It also polymerizes at mild temperatures and pressures, making it an attractive plastic for industrial use due to its ease of manufacturing. For these reasons, polyethylene has emerged as a target contaminant for microplastic studies in the environment.

Additionally, microplastics have a widespread geographic distribution given their diversity in origin (industrial facilities, water systems, litter deposits). Therefore, a multidisciplinary approach to their observation and quantification is necessary in attempting to understand the scope of their contamination and subsequent environmental effects.

HEALTH IMPLICATIONS FROM PACKAGING PLASTIC MATERIALS

Different types of plastics materials are used as containers for pharmaceuticals e.g. intravenous infusions, for food packaging, and others form of packaging. Plastic containers are now extensively used in almost all places across the world. In addition to pharmaceuticals, they are used to keep household products, drinking water, food etc. Different types of toxic materials are known to leach out of different types of plastics. In a previous analysis study of deionized water kept in different types of plastic containers at different storage conditions was conducted.

The major observation was that water stored in all types of plastics used (PET drugs and water bottles, PVC, PP and PE) at all conditions (freezer, stability chamber, refrigerator, room temperature and sun light) was found to exhibit U V absorption. This is a clear indication that some U V absorbing materials leached out. Sun light seems to affect the process of leaching since storage under direct sun light resulted in water turning acidic or with oxidisable substances or with ammonium, depending on the type of plastic used. It is recommended that materials stored in plastic containers especially PVC should be stored away from direct sun light.

Higher than normal lead content was detected in water stored in all plastic containers (except drug bottles) kept in the refrigerator. Water stored in PE showed a higher concentration of lead (90μg/L) when stored in the refrigerator and even a higher concentration (190μg/L) when stored under direct sun light. Utmost care should be taken in choosing packaging plastic containers to store water and food, since lead is known to be toxic.

Result of analysis of the plastic containers revealed that all types tested (except PP infusion bottles) could not be sterilized. They showed physical changes upon sterilization. Sterilization by heat resulted in deviations from specifications in some tests e.g. reducing substances in PET drug bottles.

According to World Health Organization (2011) report, care should be taken in selection of plastic type and storage condition, if plastic containers are to be used for pharmaceuticals or food. It is recommended to avoid using certain types of plastics, particularly PE as a container for food or water.

Plastic Materials Composition

Plastic is a polymer which is a Greek word (poly= many, mer= part). Polymers are substances composed of molecules (monomer) containing one or more species of atoms linked to each other covalently. For example, polyethylene which is the material for packaging plastic is derived from natural gas or a fraction of crude oil. There are different types of polymer classifications. These include:

Natural polymers: They include DNA (deoxyribonucleic acid), RNA (ribonucleic acid), proteins, and cellulose. Different inventors and scientists started converting some of these natural polymers into useful products in the 18th century.

Synthetic polymers: The production of these types of polymers started in 1909. Scientists discovered if these polymers are cross-linked into long chains, the resulting products are usually thermoset (resistant to heat) and more rigid. Thermoplastic polymers can be heated and remoulded into many different items. Some of the synthetic polymers include polythene, polyvinylchloride, nylon, polystyrene, and polyesters. These polymers are mainly used to make some of the commonly used plastic packaging materials (bags and containers).

Addition polymers: These are formed from alkenes. They are made from units (monomers) which are made of carbon and hydrogen atoms. Oxygen, halogen, and nitrogen atoms can be represented. Typical monomers are ethylene (CH2 =CH2) such as vinylchloride (CH2 =CHCL). These are linked together without loss of atoms. Monomers have C=C bonds which open up as the monomers join in chains. Ethylene molecules can join together under the proper condition of temperature, pressure and catalyst to form polyethylene.

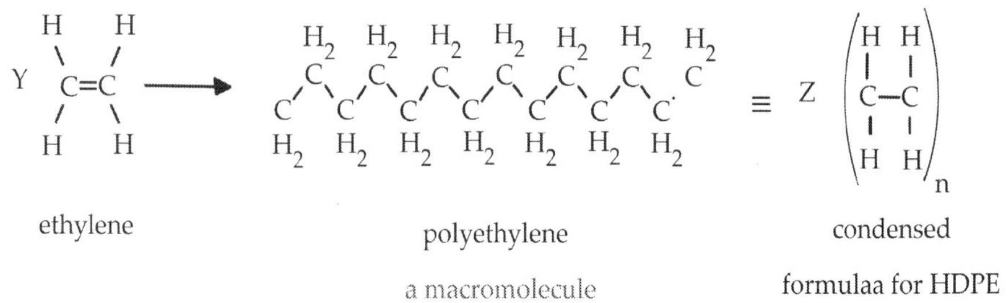

Source:https://www2.chemistry.msu.edu/faculty/reusch/virttxtjml/polymers.htm

Figure 6.3: A typical polyethylene production process (High-density polyethylene).

This process of polymerization reaction is usually initiated in the presence of substances that have an unpaired electron (free radical). In other words, a reaction between the double bond and free radical usually occur to form a long chain of polymer.

Some polymerization reactions always occur by means of carbonium ion which acts as a catalyst that helps in the generation of the following;

Condensation polymers: These polymers are usually formed from the amines, carboxylic acids, acid chlorides, and aromatic compounds. They are formed by reaction of monomers. The different between this reaction and other polymerization reaction is the elimination of small molecules such as hydrochloric acid or water.

Uses of Polymers

Polyethylene is essentially a mixture of very long chain of alkanes that makes it useful in synthesis of a variety of plastic materials such as plastic containers, plastic bags, and other plastic materials. In synthesis of plastic materials, ethylene is usually mixed with other substances such as additives, co-monomers, solvents, and other chemicals then polymerized to give the polymer which is extruded, pelletized, or flocked to give a product that is commonly referred to as resin. Resin is then used to make plastic containers, plastic films/bags, and other different products. According to studies, it has been proven that six resins types are used to make more than 90% of plastic packaging materials.

The following table illustrates some of the safe plastic materials and those that should be avoided in day to day activities.

Table: 6.1: Illustrating some of the safe and unsafe plastic products to use and to avoid

Safe plastics	Plastics to be avoided
Polyethylene: Absorption of water for this type of plastic is usually low. Chemical, solvent, and electrical resistance of polyethylene is also excellent. • They are widely used in pharmaceutical industry and in hospitals as containers for liquid or dry products, sterile devices packaging and to moulded parts for a variety of devices and equipment. • Some of the commonly known polyethylene materials include high-density polyethylene, low-density polyethylene, terephthalate, and the polyethylene. ***Polypropylene:*** They could be used in every application for which polyethylene is used. Devices made of this type could be autoclaved for sterilization.	***Polyvinyl:*** One of well know polyvinyl is referred to as vinyl chloride. This is the second most commonly used plastic in the world. Other polyvinyl plastic products include vinyl acetate and vinylidene chloride. ***Polystyrene:*** They are mainly used for fabrication of packaging plastic containers. They have low heat resistance. ***Polycarbonate:*** They are strong and have high temperature resistance. They have hardness properties similar to those of metals and can replace metals in numerous industrial applications. Some carcinogenic effects occur due to additives added during manufacturing of these types of plastic products. Such substances can leach. Occasionally they have been the main causes of adverse health impacts such as haemolysis of blood cells, thrombosis, hypersensitivity reactions, precancerous changes, and local tissue necrosis. • These are used to prepare materials ranging from soft flexible packing plastics, to rigid hard packaging containers. They are characterized by high temperature resistance with softening points (70 -180 °C or higher), high solvent and chemical resistance, low water absorption and moisture permeability and non-flammability. • They are used in pharmacy and medicine fields as in the manufacture of blood bags and intravenous solution packaging containers. • The unplasticized form is used in fabrication of rigid parts for devices and administration equipment in hospital.

Source:http://www.everydayroadtohealthy.com/2010/12/08/110/

Most plastic packaging containers such as water and soft drink bottles are usually manufactured using polyethylene terephthalate. These plastic products are meant or anticipated to be for single-use purposes. However, most of them are reuse by people for different purposes. Studies have shown that plastic breakdown is accelerated through repeatedly washing and this results to leach of various dangerous and harmful substances into packed or stored products. For instance, one of these substances is diethylhexyl adipate (plasticizer) which causes cancer in humans. It is important for the people to be enlightened about the dangers involved in reusing various plastic containers.

Uses of Plastic

Plastic materials are widely used in different sectors due to their flexibility and due to the two definite disadvantages of glass which are fragility and weight. Significant uses of plastic in products packaging have been made. For example plastics are being used in food packaging, soft drinks packaging, medicine packaging and other type of packing services.

Other uses which would have never been technically feasible without use of plastic are in dwelling catheters, prosthetic devices, tracheotomy tubes, flexible blood collection containers, and semi rigid and flexible containers for intravenous, irrigation, and inhalation solutions.

An additional use is in secondary packaging (packaging that is not in direct contact with the product itself). For these uses in products packaging, plastic safety and adequacy have actually raised many questions.

In case of plastic used in direct contact with a product the length of time that the packed product and container are in contact may determine whether problems such as discoloration, leaching, and absorption or adsorption of a constituent of the product may arise. Both the product and the package containing it could change significantly from time of manufacture. Other factors that may affect the plastic packaging and/or product are:

- Product storage conditions
- pH

- Temperature
- Type of polymer used
- Method of package preparation
- Light transmission and sterilization process

All plastic packaging containers are usually manufactured using materials that may include other substance such as additives. Health experts have recommended that these plastic packaging materials should not include other substances such as additives that can be extracted by the packed content. In other words, the packed content should not be absorbed into or absorbed onto the plastic containers.

Some of the commonly used polymers in the manufacturing of plastic packing materials include polyethylene vinyl acetate, polypropylene, polystyrene, polyethylene terephthalate, polyvinyl chloride, high-density polyethylene and low-density-polyethylene.

The nature of the additives depends on the polymer composition and the method of manufacture of the plastic container. Each source will have its own range of additives and processing conditions.

Interaction of the packaged product with the plastic container may involve permeation, leaching, absorption, chemical reaction, or alteration of physical characteristics of the polymer or the product. Permeation of gases or water vapour through polymer has a marked effect on the stability of the packaged product. If a product that contains volatile ingredients is stored in plastic containers, loss of volatile components may occur. Leaching of additives into the dosage form causes product contamination when the dosage form is in solution. Factors that affect absorption or adsorption of drug components to the polymer layer include; chemical structure, pH, concentration of ingredients, area and time of contact and temperature.

Table 6.2: Illustrating some of the key additives used in the manufacturing of packaging plastic products

Types of additives	Descriptions
Lubricants	• Are added during moulding and extrusion operation. Example is zinc stearate in case of polyethylene
Stabilizers	• Are added to prevent degradation by heat and light and to improve its aging characters. But they have many problems: Some have tendency to migrate to the surface of the moulded part during storage. Others have limited solubility in aqueous media and could be extracted into products. • Examples include: organometallic compounds, fatty acid salts and inorganic oxides.
Plasticizers	• They are used for softness and flexibility • They are used in vinyl, celluloses and propionates, and can easily migrate to the surface of the polymer and then get extracted by the product. Therefore they should be carefully selected. • Examples include: butyl benzyl phthalate (BBP), di-n-butyl phthalate (DBP), di-n-pentyl phthalate (DPP) and di-hexyl phthalate (DHP).
Antioxidants	• These are special types of stabilizers used in retarding oxidation. • They can migrate under certain environmental conditions. For instance, their combination with other additives may lead to undesirable chemical reactions. • Example includes tris (nonylphenyl) phosphite (TNPP).
Antistatic agents	• They are added to prevent the build-ups of static charges on the plastic surface. • Examples include epoxy ethane amine type and glycerol stearate
Slip agents	• These are primarily added to polyolefin (polyethylene and polypropylene) to reduce the coefficient of friction. At the end they result in anti-tack and anti-block characteristics. • Examples include: oleyl palmitamide, stearyl stearamide and faty acide amide (eurcamide).

Dyes and pigments	• They are added as colouring agents. They may also leach into product. • Due to the undesirable effects of combination and migration of additives, the final product and package should be evaluated for safety and stability under various time and storage conditions simulating those the product is expected to be subjected to. Evaluation should include compatibility of the primary plastic container with its secondary packaging. • Prolonged exposure to ultra violet light enhances the migration of certain additives and accelerates the aging characteristics of the plastic and decrease the shelf life of the product. Examples include: lead chromate and lead molybdate.

Source:https://www.aranca.com/knowledge-library/articles/ip-research/innovative-plastics-polymer-additives-for-smart-packaging

Some of Adverse Effects of Plastics on Human Health

Plastics used in packaging may cause various adverse effects either by direct contact with tissue or indirectly via solutions stored in plastic packaging containers. Adverse effects may also rise when workers are exposed to plastic by inhaling fumes during manufacturing or by handling the materials. Pure polymer plastics usual appear of low toxicity, but some carcinogenic effects occur due to prolonged implantation. Substances added during manufacturing to impart specific physical properties are toxic such as plasticizers(phthalates). According to Co-op America, (2004-2005) report, such substances can leach and have adverse effects on human beings such as;

- Haemolysis of blood cells
- Thrombosis
- Hypersensitivity reactions
- Precancerous changes
- Local tissue necrosis
- Developmental and reproductive damage

Incineration of polyvinylchloride leads to emission of dioxins. These substances are known for their dangerous health impacts on human beings. Dioxins are known to cause immune challenges on human beings, reproductive challenges, and different types of cancers. For instance, according to a recent study that was conducted by United States Environmental Protection Agency (US EPA) revealed that human beings can develop certain nervous system challenges such as nausea, body weakness, and loss of concentration when they are exposed to styrene for a short period of time. On the other hand, when exposed for a prolonged period of time to styrene, they are likely to develop adverse health effects such as cancers, adverse damage on nervous system, and various liver complications.

According to Environment Health Perceptive (2005) report, it was discovered that different polycarbonates in plastic products can actually leach bisphenol-A, which is a compound known to cause disruption of endocrine system in human beings. Some of the main effects of this substance include alteration of brain chemistry, altering reproduction system of both males and females, changing the structural and behavioural components in human beings among other effects.

Food and water plastic containers may be a potential source of unwanted and possibly harmful chemicals. Antioxidants are often used to preserve the durability and aesthetic quality of these plastic containers. A popular antioxidant found in plastics is called tris (nonylphenyl) phosphite (TNPP) which contains three units of nonylphenol per molecule of this agent. When introduced several years ago, these materials were considered inert. Since the recent discovery of their potential to be endocrine disrupters, concern over human exposure to this chemical has increased. Stabilizers (organometallic compounds, fatty acids, and inorganic oxides) are some of well-known additives with the greatest potential of toxicity. When these substances are used together with other substances such as plasticizers in manufacturing of packaging plastic products, the toxicity of the plastics is usually increased greatly.

STABILITY OF PACKAGED PRODUCTS IN PLASTIC CONTAINERS

All products packaged in plastic containers should be stable. The products should be safe, officious, and acceptable to the end users. In other words, product stability in plastic packaging containers is usually associated with its resistance to biological reactions, physical interference, and chemical changes. This means that the packaged product's original properties should not be altered in any way, especially during transportation of the products, storage of the products, and use of the products. Quantitatively, stability is the shelf-life of packaged products in its closed plastic container under specific conditions and can be defined as "the time from manufacture or preparation until the original potency or content of active ingredient has been reduced by 10%." More stringent limits may be required, especially if the degradation products are more toxic or irritant. There are numerous reactions that may actually result into degradation of the packaged products in plastic containers. Some of these reactions include; polymerization, isomerization, dehydration, hydrolysis, and oxidation.

Many plastic packaged products may degrade when exposed to light (photochemical reactions). Some of the factors that may affect the stability of packed product in plastics include pH, nature of the product, general acids and bases, ionic strength and product's concentration. Scientifically, plastic packaging containers should not interact physically or chemically with the formulation so as to alter the strength, quality, or purity of its contents.

The choice of plastic packaging container and its closure may affect the stability of its content. Some of the packaging elements are subject to physical and chemical changes which may be time and temperature dependant. Some of these elements include;

- The appearance of plastic container with special emphasis on inner walls
- The migration of ingredients onto the plastic or into rubber closure
- The migration of plasticizer or other components into the formulation
- The possibility of two-way moisture penetration through the container walls
- The integrity of the seal
- The back off torque of the cap

Currently, different health experts are concerned with plastic containers so as to evaluate their stability and their effect on stability of the contents (deionized and distilled water).

For instance, plastic containers are most popularly used for storing pharmaceutical products. Selection of the type of the polymer to be used in synthesis of container may affect the suitability of plastics for pharmaceutical use. Many tests are used e.g. biological procedures for determining the suitability of plastic for packaging products intended for parenteral use. Materials from the plastic can leach into the formulation, and materials from formulation can be absorbed onto, into, or through the container wall. For instances;

- Volatile oils, flavouring and perfume agents are permeable through plastic to varying degrees
- Components of emulsions and creams have been reported to migrate through the walls of some plastic packaging containers
- Loss of moisture from formulation
- Gases like oxygen, carbon dioxide may migrate through the container wall from the air to affect the preparation
- Solid products, when packaged and stored in plastic container may be affected by moisture penetration
- Closures must also be studied in stability. Closures must form an effective seal for container; they must not react physically or chemically with the packaged or stored content
- Absorb materials from the formulation or leach ingredients into the content.

EVALUATING ENVIRONMENTAL HEALTH PROBLEMS OF PLASTIC PACKAGING PRODUCTS

As stated before, the length of time a packaging plastic container and a product are in direct contact may determine problems such as leaching. Many other factors that may affect plastic packaging and products include storage condition, pH, temperature, type of polymer used and sterilization process. Substances added to plastics during manufacturing for specific physical properties are toxic. Such substances can leach causing many adverse effects like haemolysis of blood cells, thrombosis, hypersensitivity reactions, precancerous changes, local tissue necrosis and developmental and reproductive damage.

According to recent review studies by Environment Health Perspectives (2005) regarding bisphenol-A's (BPA) effects have found that more than 80% of published studies reported effects of various substances that are normally added in packaging plastic containers including alteration to brain chemistry, structure, behaviour, the immune system and male and female reproductive systems. Examples of such added substances are:

Phthalate: They are added to impart flexibility and durability to plastic. They are not covalently bonded and leach when in contact. They may also be released into environment during production and use

Antioxidants: They are added for durability and aesthetic quality. In previous research studies, nonyl phenol (NP) was found in water from HDPE and PVC containers

Acetaldehyde: When it leaches and diffuses in the product, it causes an off taste in bottled water (10-20 parts/ billion). Freezing water bottles may also cause plastic to leach into water.

Many of these substances may not cause harm individually, if they occur in low concentration, but the combination effects from exposure to all these substance may cause very adverse health impacts to human beings, even at low concentrations, if there is a prolonged exposure.

The diversity, abundance, and hazardousness of various substance added during manufacturing of packaging plastic materials may contributed to very severe environmental challenges. For instance, during plastic products life cycle, numerous hazardous substances may be emitted in to environment which may cause different life threatening problems to human beings and other living organisms. Therefore, it is the responsibility of the national governments and other non-governmental agencies to consider the large and growing global consumption of plastic packaging products, and their omnipresence and persistence in the environment. Assessing various health risks and hazards of these large material groups is of great benefit.

However, different people have different perspectives on the use of plastics in the society. For instance, majority of people consider plastic materials to be of greatest benefits over other materials because of one reason or another. For instance;

Plastic packaging products usually help different people in protecting their goods and food from being contaminated and from getting wasted. This is considered as one way of conserving resources.

Compared to other packaging products, plastic products are usually light-weight, thus save fuel and decreases emissions during transportation.

- Plastic water supply systems and storage containers/tanks provide clean water
- Light plastic materials (replacing metals) in cars and aircraft save fuel and decreases emissions.
- Efficient plastic insulation materials in buildings save energy and provide climate protection
- Plastic protective clothing and safety equipment (e.g. fire proof materials, helmets, air bags) protects form injury
- Solar heaters and solar panels, in parts made of plastics, provide renewable energy.

However, the extent of this plastic product pollution, especially by micro-plastics, is much more prevalent than it is thought by many people, both in terms of greater quantities and smaller particles.

Studies have shown that various polymers (non-biodegradable) are usually degraded through a number of processes such as ionic radiation, oxidation, heat and sun light exposure, mechanical shearing, and hydrolysis. Additionally, it has also been revealed that degradation of these non-biodegradable polymers can also occur in presence of other environmental pollutants such as sulphur dioxide, carbon monoxide, ozone, and nitrogen oxide. These processes usually cause polymers to get brittle, to fragment into small pieces and to release degradation products. Different degradation mechanisms exist and their dominance mainly depends on the type of the polymer. There is a chain of scission that is involved during the breakdown of the chemical bonds in various polymer molecules. This breakdown process is often random, but for some polymers it precedes at the polymer end chains and the initial monomers are broken off, a process called depolymerisation. Since plastic products are composed of many different chemical substances, and the main part of these substance is usually broken down into something completely different, prediction of fatality of these substances may complicated.

Human Exposure to Plastic Hazard Substance

Consumers exposure to chemicals released from a plastic material during the use phase may vary greatly between different products. The exposure may for instance come from indoor air, food, water, and skin contact, but in most cases this is not likely to be so large that it will result in adverse effects. However, the actual exposure does not come from only one product but from a whole array of different plastic products, and exposure to a mixture of chemicals will often be continuous.

Traces of chemicals that are used in manufacturing of plastic products (such as bisphenol A, phthalates, and brominated flame retardants) have been identified in human beings.Some consumer categories are particularly sensitive to hazardous plastic chemicals, for instance infants, and larger margins of safety precautions are therefore required.

Conducting tests on plastic products can be advantageous for screening purposes, especially when the chemical composition of the test medium is not known, and thereby not either the anticipated toxic response. Chronic tests are usually more sensitive and ecologically relevant, since exposure to toxicants in the environment usually occurs in lower concentrations during a longer time period. Examples of other toxic effects include carcinogenicity, reproductive toxicity, mutagenicity, and various effects caused by endocrine disruption.

Many of the chemicals that are used to make plastic polymers and plastic products have been tested for toxicity to a varying extent, and the assembled data have provided the basis for harmonised environmental and/or health classifications for these substances. These hazard classifications are used to identify the potential environmental and health hazards associated with each plastic polymer type.

The most frequent reported hazard classifications among the plastic polymers and monomers include skin sensitization, respiratory irritation, acute toxicity (inhaled, skin, oral), serious eye irritation, carcinogenic, and skin corrosion.

Table6.3: Plastic polymers and monomers hazards classifications

Monomers/Polymers	Hazard classifications
• 1,3-butadiene	Carcinogenicity and Mutagenicity
• 1,4-ichlorobenzene	Aquatic chronic, Carcinogenicity, Aquatic acute
• 4,4'-methylenedianiline (MDA)	Carcinogenicity, Mutagenicity, Skin sensitization, Specific target organ
• 4,4'-methylenediphenyl diisocyanate (MDI)	Respiratory sensitization and Skin sensitization Carcinogenicity, Specific target organ toxicity - repeated exposure
• Acrylamide	Carcinogenic, mutagenicity, specific target organ toxicity, repeated exposure, reproductive toxicity, and acute toxicity
• Acrylonitrile	Carcinogenicity, Skin Sensitization, Acute, Aquatic chronic toxicity, Serious eye damage
• Benzyl butyl phthalate	Reproductive toxicity , Aquatic chronic, Aquatic acute
• Bisphenol A	Skin sensitization, Reproductive toxicity, Serious eye damage
• Epichlorohydrin	Carcinogenicity, Skin sensitization, Skin corrosion, Acute toxicity
• Ethylene oxide	Carcinogenicity, Mutagenicity, Acute toxicity
• Formaldehyde	Skin sensitization, Carcinogenicity, Acute toxicity, Skin corrosion

• Hexamethylenetetramine	Skin sensitization
• Maleic anhydride	Respiratory sensitization, Skin sensitization, Skin corrosion
• Methyl methacrylate	Skin sensitization
• m-phenylenediamine	Mutagenicity, Skin sensitization, Aquatic chronic, Aquatic acute, Acute toxicity
• Phenol	Mutagenicity, Acute toxicity, Specific target organ toxicity
• Phosgene	Acute toxicity, Skin corrosion
• p-phenylenediamine	Aquatic chronic, Skin sensitization, Aquatic acute, Acute toxicity
• Propylene oxide	Carcinogenicity, Mutagenicity
• Toluene-diisocyanate (TDI)	Acute toxicity, Respiratory sensitization, Carcinogenicity
• Vinyl chloride	Carcinogenicity

Source:http://www.chem1.com/acad/webtext/states/polymers.html

Most Hazardous Ranked Polymers and Toxic Plastic Products

When comparing the polymers types which are ranked as most hazardous with the leachates from plastic products that are toxic to human beings, there is a correlation. The most hazardously ranked polymers used in plastic products include polymers polyurethane, polyacrylonitrile, polyvinyl chloride, and epoxy. These substances are also most toxic leachates, with a few exceptions.

Since the ranking is based on monomer composition this would (if this correlation is valid also for a greater number of plastics) mean that residual monomers are leached in such high quantities from plastic products to cause acute toxicity. For the epoxy products, this correlation is likely to be valid. Uncured resin DGEBPA or its monomers (bisphenol A and epichlorohydrin) and some of the curing agents are the most likely cause of toxicity seen for the epoxy products. For the other polymer types there are several other possible explanations for toxicity. The PVC and polyurethane products are soft to semi-soft, which facilitates migration of chemicals. In the plasticised PVC also the quantities of plasticisers and other non-bound additives, which may be available for leaching, are larger than for other plastic types. The polyacrylonitrile products are usually synthetic textiles. Any residual monomers in the fibres are likely to be released during textile manufacturing processes, and, besides, there are many other possible chemical residues emanating from the processing of the textile fabric that may cause toxicity. Other reasons that argue against monomers being responsible for most of the toxicity include; firstly, the residual monomer content is generally quite low in a plastic product and, therefore, possible effects are more likely to be seen in chronic studies. Secondly, many of the hazardous monomers are volatile and may be lost during the leaching.

According to European Commission (2008) report, a few of the monomers used in plastics are under evaluation for endocrine disrupting properties within the European Union strategy for endocrine disrupters. These monomers include;

- Bisphenol A (used in polycarbonate and diglycidylether of bisphenol A)
- Styrene (used in polystyrene and polystyrene copolymers)
- Epichlorohydrin (used in the main epoxy resin DGEBPA)
- Some phthalates (used in PVC)

The polymers containing these monomers could be ranked as more hazardous if endocrine disruption is identified. Another factor that can affect the ranking is a subsequent assessment of the environmental fate of the monomers in the hazardous ranked polymers. For instance, if the monomer is very volatile and/or very rapidly photodegrading, readily biodegradable, and not bioaccumulating, the hazards may mainly be confined to the production phase with worker exposure through air. This is because many of the monomers and solvents used are volatile.

SUSTAINABILITY IN PLASTIC USAGE

The risk assessment of plastic products is very important. This helps in determining various hazardous substances that may threat lives of human beings. To have sustainable plastic usage, there are a number of considerations that require emphasize. These include:

Considering the fact that plastic production globally is an omnipresent occurrence both in the environment and the society at large,

Considering the persistence and partial degradation of plastic products,

Considering the large amounts of plastic waste and microplastic pollution in the environment,

Considering plastic diversity by means of types and applications,

Considering littering, poor waste management and low recycling rates,

Considering recycling obstacles because of heterogeneity, as well as potential emissions of hazardous chemicals during the entire lifecycle

Exposure and effect assessments are important in order to assess the risks associated with the use of plastic products. This can be done by predictions of environmental fate for the monomers, and with data on degradation and degradation products of plastic materials under environmental conditions. Long term leaching, which reflects specific exposure scenarios during the use phase and the end of life phase under the influence of degradation would be valuable, and bioassays studying chronic effects would be desirable. Research on occurrence of plastics in environment and the effects of microplastics is on-going. In conclusion, there is too little knowledge of the release of chemicals from plastic products and potential effects.

According to quotation from American Chemistry Council (2011), "In today's world, life without plastics is incomprehensible". That is very true, plastics are extremely necessary in modern society, but some of the current uses and misuses are not sustainable. A more sustainable use of plastics can be achieved by many measures. There is a large potential for reducing the per capita use in the industrialized world, for instance by eliminating excessive packaging material and practicing material reduction whenever possible. By developing more plastics based on bio-based raw materials, the great dependence on non-renewable crude oil and the environmental consequences of crude oil extraction and refining, as well as the contribution to global warming during the end of life phase, can be decreased. Persistence is desired in many applications for long term use, but for short term and single use applications, especially in the packaging sector, the use and further development of biodegradable plastic materials that are fully degradable in the natural environment, are important. Recycling of plastics involve problems since a homogeneous fraction is needed for a similar grade end-product, but the recycling could be facilitated by practicing Eco-design for recyclability and systems for collecting recyclable fractions. Increased recovery (mechanical, chemical or energy recovery) when possible, saves resources and decreases the area needed for landfill. Global action and strong measures to reduce littering are essential to protect the environment. There are a number of strategies that can be put in place in order to have sustainability in plastic usage, especially in packaging sectors. These strategies are discussed below.

Product Take-back Programs

Take-back policies require producers to collect or accept returned plastic materials from consumer after use and arrange for their appropriate treatment or disposal. Take-back policies can be legislated by government or may be initiated voluntarily by producers. Take-back programs place the responsibility on producers to manage financially and physically the waste generated by their products and packaging, thus stimulating the efficient use of inputs, the design of ecological sensitive products and a reduction in the amount of plastic waste materials being discarded in landfills.

It may not always be physically feasible or cost effective for a producer to collect and treat its own products. Most legislation allows formation of a third party organization, or Producer Responsibility Organization (PRO), to oversee the producers' waste management obligations. Producer Responsibility Organizations finance their operations by charging members fees. For these organizations to be effective at attaining goals and objectives in plastic waste management, fees must reflect the actual cost to collect, sort and treat the materials. This can be accomplished by basing fees on the weight, volume and package composition, thereby rewarding companies designing goods which contain less material or are easy to recycle.

Economic Instruments

Economic instruments are designed to encourage environmentally efficient products and purchasing decisions by providing direct financial incentive or disincentive to one or more actors in the product chain. Research into the effects of fiscal incentives for changing environmental related behavior has identified a number of advantages and disadvantages of such methods.

Economic incentives tend to be more cost-effective than the commonly used command and control (CAC) approaches. Command and control regulations achieve government goals by defining how businesses must conduct certain aspects of their operations. These types of regulations are commonly used to achieve pollution control objectives, dictating to firms "how much pollutants they may emit, the type of technology to use, the goods they may produce, the production processes to employ, etc.". Economic tools, on the other hand, allow producers to decide the most effective and efficient way to alter their behavior and reduce their ecological impact. If properly applied, fiscal tools may encourage producers to continually improve their practices by providing increased financial benefits for acting in an approved manner. With CAC approaches, once the environmental standard is reached, there is no longer incentive to reduce their impact any further. Furthermore, studies have found that economic instruments tend to achieve results more rapidly and involve less administrative expenditure.

However, the potential outcomes resulting from the use of economic mechanisms are often complicated to predict, for it is difficult to estimate how consumers and industry will react to new market signals. Unfortunately, the incentives do not always produce the intended behavioral changes. For instance, many governments employing landfill taxes have had problems with the illegal dumping of waste. Furthermore, governments do not always feel comfortable relying on market mechanisms to achieve their policies. Fiscal instruments allow business great decision-making power and result in less government control. Finally, economic tools result in additional costs for businesses and can impact competitiveness if not all producers in an industry is subject to the same taxes or fees. Typically within a province or nation fiscal instruments will uniformly impact all producers in a sector. However, producers involved in trade beyond the borders of the regulation may find more challenging. The most commonly used economic tools are taxes, deposit-refund programs and advance disposal fees.

Taxes

Taxation is a commonly used economic tool and can readily be incorporated into the existing business infrastructure. Taxes evoke different responses when applied to different stages in a product's life cycle. Taxes can be applied to ecologically damaging resource extraction or production processes, non-renewable raw materials, virgin or toxic materials, unsound production process, non-recyclable products, waste collection, and landfill usage. There are several types of taxes that can be established with regard to management of plastic waste materials within a country. These include;

Virgin Material Taxes: The intent of material taxes levied on virgin inputs is to reduce the amount of plastic materials utilized in products/packaging, to decrease reliance on virgin materials, to encourage the use of recycled inputs, and to further the development of new markets for goods manufactured from secondary materials. When the revenue generated from taxes is set aside for reuse and recycling programs, plastic waste reduction goals can also be achieved.

Alternatively, some governments have chosen to reduce subsidies provided to raw resource extraction and processing activities. The removal of subsidies forces resource extractors to increase the costs of virgin raw materials, bringing them more into line with secondary materials.

Product Taxes: Some governments have levied a tax on plastic products that have a negative impact on the ecosystem. Such taxes serve multiple purposes, such as deterring the production of such goods, correcting externalities (internalization of costs), and generating revenue. Problems may arise with product taxes when revenue generated is entered into general government revenue, rather than being earmarked for recycling or treatment programs. In such cases, funds are not being used to reduce plastic waste pollution, and producers and consumers become resentful of the tax, viewing it as a 'money grab' rather than an environmental initiative. Other problems arise when taxes are levied on essential products. Consumers in lower tax brackets may suffer as a result of the tax, unable to afford the product.

Waste Collection Taxes: Currently, most households around the world do not pay directly for their waste collection and treatment; rather, households pay for these services through municipal taxes. As a result of indirect fees, consumers have no incentive to reduce their garbage production.

Some municipalities in a number of countries have recently adopted, or are considering new payment schemes for plastic waste management in which households are charged a tax or fee based on the volume or weight of the garbage they produce. While such systems send a signal to households to be more conscientious consumers and minimize waste production, waste collection charges do not provide an incentive to producers to alter their material selection practices, product designs or production processes.

Deposit-Refund Programs: Under a deposit-refund program, the consumer pays a fee (deposit), when purchasing certain plastic products, which is fully or partially refunded when the product is returned to the retailer, producer or appropriate treatment facility. When plastic goods are returned to the producer, either directly or via the retailer, the producer assume financial and physical responsibility for the proper treatment of the product, thereby encouraging the producer to use less material and improve product design in order to reduce their financial and administrative obligations.

Deposit-refund programs provide financial incentives to consumers to return waste materials, leading to significant increases in material recovery rates. In addition to increasing the amount of material diverted from landfills, these programs also reduce litter and ensure a reliable source of post-consumer materials for companies manufacturing products with recycled inputs.

Source:https://www.packagingnews.co.uk/news/swedens-deposit-return-scheme-motivates-scotland-28-08-2013

Figure 6.4: An illustration of deposit-refund program for plastic waste containers.

To encourage reuse over recycling, some governments have combined their deposit refund system with an environmental levy. However, critics state deposit refund programs often encourage the use of environmentally damaging products (i.e. those products which are not a part of the deposit-refund initiative because they cannot be reused or recycled). Those opposed believe that consumers will be encouraged to purchase these unfriendly products because they will not have to pay the deposit and will not have to return the plastic containers. Advocates believe this problem can be overcome by either levying a tax on containers which do not fall within the deposit-refund scheme, or ensuring that policies apply uniformly to all products within a particular category regardless of whether or not they can be reused or recycled. Another shortcoming of deposit refund schemes is the extensive and costly infrastructure (e.g. administrative system, enforcement program, collection and storage facilities, education campaigns) required to support these systems.

Furthermore, these programs remove containers that have value and are easy to process from the recycling stream, such as PET bottles, increasing the cost of operating municipal recycling program.

Advance Disposal Fees: Advance disposal fees (ADF) are levies placed on certain products, which are paid by consumers at the point of purchase. The revenue from these fees is used to pay for local recycling initiatives. For an ADF program to be effective in plastic waste reduction there must be an agreement or regulatory provision stating producers will assume full or partial physical responsibility for their used products.

Advance disposal fees programs are useful for achieving plastic wastes reduction targets, but do not tend to meet source reduction and advance disposal fees goals. This is because the consumer pays the levy. If the cost is not borne by the producer, there is no financial incentive to reduce their inputs or redesign their products. To combat this deficiency, some governments relying on ADF programs charge a reduced levy for products in which plastic waste management costs have decreased as a result of improved design (e.g. contain less material, increased recyclability. It is believed consumers will seek out products with lower levies, thereby providing a competitive advantage to producers of 'green' products. Advance disposal fees initiatives will be more effective if delivered in conjunction with an education program to inform consumers of the differences in levies and the types of products with lower fees.

Governmental Instruments

Aside from legislating economic instruments or take-back programs, there are a number of other measures governments can implement to their objectives in plastic waste reduction.

Targets and Quotas: Targets and quotas assist policy-makers to gauge the success of an initiative and to continually improve the efficiency and effectiveness of programs. Measurable objectives ensure performance expectations are clear, and compel responsible parties to take action. Many governments have set targets or quotas for reuse, recovery, and recycling, as well as minimum- recycled content. Quantitative performance measures are most successful when legislated and supported by enforcement provisions (e.g. threats of further regulation, fines).

For instance, recovery and recycling targets is one of such initiatives. In 1994, the European Union devised Directive 94/62/EC on plastic packaging and packaging waste, which established recovery and recycling targets to be met by producers in each member country for primary, secondary and tertiary packaging materials. The targets that were to be met by June 30, 2001 include overall *Recovery Rate* (50 to 65-percent by weight) and *Overall Recycling Rate* (25 to 45-percent by weight).

The data collected found that all but three of the participating members had reached the recovery target and all but one had achieved the recycling targets. The EU Directive is currently under review and new, more ambitious, quotas have been proposed.

Material, Product and Landfill Bans: Some governments have implemented legislation banning the sale of plastic products or materials deemed to be pollutants or hazardous to human health. Bans can also be used to restrict the disposal of plastic products. Landfill bans can be placed on plastic products that the government has identified as hazardous and may leach into the soil, or on recyclable products, to ensure they enter the proper treatment stream.

Bans should be used with caution because replacing plastic packaging materials may result in unforeseen problems. For example, new or substitutes products may consume more energy in production and/or use, be of substandard quality, or be significantly higher in price than banned items.

Bans set by municipalities may cause problems for producers who supply their product nationally. Critics of local product/material bans cite that such regulations may be used as inter-provincial trade barriers, restricting the import of plastic products from other places in order to favor local suppliers. Many producers are also concerned that if a government introduced unique product bans, it could increase their costs considerably, especially if it required them to invest in new equipment, or if they had to supply different packaging to different jurisdictions.

Government procurement policies: Governments have considerable purchasing power, which may be used to influence the plastic design and material selection decisions of manufacturers. Green purchasing policies direct government agencies to purchase eco-friendly products (e.g. reusable, recyclable, non-toxic, recycled-content, less materials). For example, a policy may state that preference be given to environmentally compatible products as long as the price differential is no more than 10 percent.

Green purchasing policies inform suppliers of the government's preferences for products with a negligible impact on the environment and in doing so, can aid in the development of markets for goods designed in a benign manner, and set a positive example for the business sector. Unfortunately, green products tend to be higher priced, thus cost is a limiting factor for government purchasing initiatives. However, as government departments and agencies at the federal, provincial and municipal levels get involved in eco-friendly procurement, the cost of green products will likely decrease.

Eco-Labeling Programs: Some countries have also taken strides to encourage the development of eco-friendly products. One such initiative has been eco- or green labeling programs. Under such programs, producers seek permission to use logos or seals on their goods to signify to the public that their products meet certain environmental criteria. Some manufacturers have established their own internal programs, meaning they devise their own environmental criteria and place their corporate eco-label on their products. Other producers seek accreditation by a recognized assessment organization, such as National Environmental Management Agency (NEMA) in Kenya. These third-party organizations are responsible for establishing the environmental standards that products and services must meet. If a producer for plastic packaging materials meets the criteria, it is granted the right to use the organization's environmental logo.

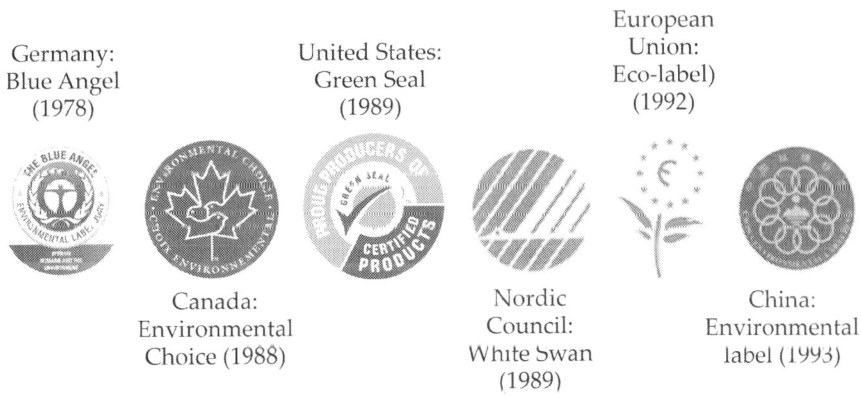

Source: http://slideplayer.com/slide/4961603/

Figure 6.5: Some of typical Eco-Labeling Programs.

The goal of eco-labeling programs is to aid the consumer in identifying products on the market that are friendlier to the environment, thus allowing them to make well informed purchasing decisions. These programs are also beneficial to producers and distributors of green goods, as eco-logos serve as a marketing tool to promote their products.

Critics of eco-labeling programs state there are simply too many logos on the market and many labels promoting false claims. They believe the environmental criteria set are often not stringent enough and many assertions are never substantiated. Other critics have raised concerns over the use of eco-labeling programs as trade barriers. Under the WTO guidelines, a nation is not allowed to discriminate among products on the basis of production processes. Advocates, on the other hand, argue that eco-labeling enhances free and fair trade by ensuring demand is driven by a well-informed consuming public.

7 FORMAL AND INFORMAL ACTORS IN MANAGEMENT OF PLASTIC WASTE MATERIALS USED IN PACKAGING INDUSTRY

INTRODUCTION

Industrialization and globalization have increased the quantity and quality of goods that are produced and moved around the world mainly through trade. This has led to an increased generation of plastic waste materials used in packaging of these goods. Plastic waste materials are often discarded with no real attachment or need for repair, reuse or recycling, leaving some developing countries in a dilemma on how to handle the increasing waste given their weak economies (and thus low technical capacities and poor physical infrastructures), inabilities to enforce environmental legislation, financial mismanagement and poor administrative capacities.

In recent years, the volume of plastic waste has been increasing at an alarming rate posing formidable challenge to communities and governments. The complexities and magnitude of the challenges become evident when considering other waste types to be managed that include industrial solid waste, municipal wastewater, storm water and hazardous waste. According to UNEP (2005) report, different government agencies are mandated to manage different waste sectors. This fragmented approach to waste management, coupled with a lack of clear definition and delineation of the different waste types, makes current waste management practices in most countries difficult.

Lack of interest and know-how on the part of the government, coupled with many sources and types of plastic waste generated from different settings and of different nature, calls for a common attention and operations strategy in plastic waste management.

Management of plastic solid waste materials is a complex task which depends on organization and cooperation between households, communities, private enterprises and municipal authorities as much as it does upon the selection and application of appropriate technical solutions for plastic waste collection, transfer, recycling and disposal.

Plastic waste management encompasses the functions of collection, transfer, resource recovery, recycling, treatment and disposal of waste. The primary targets of management of plastic waste materials is to protect the health of the population, promote environmental quality, develop sustainability, and provide support to economic productivity.

To meet these goals, sustainable plastic waste management systems must be embraced fully by local authorities in collaboration with both the public and private sectors. Although in developing countries the quantity of plastic wastes generated in urban areas are low compared to industrialized countries, proper management systems still remains inadequate. A significant portion of the population worldwide does not have access to waste collection service thus, only a fraction of the generated waste is actually collected. Systems for transfer, recycling and/or disposal of plastic waste are unsatisfactory from the environmental, economic and financial points of view.

In a country like Ghana and other several African countries, local authorities are charged with the responsibility of collecting and disposing of solid municipal wastes within their areas of jurisdiction. At present, management of solid waste in most of these countries is disposed in open dumps that lack proper environmental pollution control and monitoring mechanisms to mitigate the effects of the pollution.

UNDERSTANDING THE INFORMAL SECTORS IN PLASTIC WASTE MANAGEMENT

Informal activity may be referred to as behaviour that is extra-legal or extra-regulatory. These definitions escape the negative connotations associated with the terms illegal, shadow or underground-terms that may be too harsh when used in reference to the informal sector of less developed nations.

The nature of the informal sector depends upon the political and geographical context. In an economically advanced state with a reasonably well- functioning government that is responsive to the needs of its citizens, informal activity often refers to criminal or predatory activity among persons who knowingly and deliberately seek to circumvent state control. In this scenario, informal transactions deprive the state of tax revenue and information and render customers absent of state protection. In a poor state with a relatively weak government it would be wrong to label the entire range of informal activity as criminal or predatory. Where a state is incapable of efficiently regulating the marketplace and providing for the most basic needs of its citizens, the informal sector fulfils many important functions that the state neglects. In many countries, the informal sector is the primary source providing citizens with food, shelter, and transportation services. In such a situation, the size of the informal sector as well as the range of products and services it offers is typically much larger. Today, the informal labour market provides jobs for one-half to three-quarters of all non- agricultural employees in developing countries. It comprises 65% of all non- agricultural employment in Asia; 51% in Latin America; 48% in North Africa; and 72% in Sub-Saharan Africa.

One important exception to this is activity that occurs within the domestic sphere (e.g. childcare, housework, tutoring, etc.), which may also be referred to as informal economic activity. In most developing countries such as Africa and some parts of Asian region, the informal sector provides employment to thousands of people who would otherwise have little or no income. Given the dearth of formal wage employment opportunities in parts of the world, the informal sector constitutes a very important and dynamic part of a country's economy. It provides the bulk of urban transport services, as well as housing and food for the urban poor. In a country like Ghana, the informal sector has an estimated national employment growth rate of about 5.6% per annum compared with less than 1% in the formal sector.

For the last few decades, informal sectors in management of plastic and other solid waste materials have really grown in different parts of the world. Basically, most of informal sectors are usually not managed or regulated by the national government. It is also worth noting that informal sectors in plastic waste management are not grounded or sheltered by the formal institutional matrix of a larger social unit like a nation state.

According to World Bank (2001) review, informal sector is composed of economic activities that are not, but are supposed to be, declared to the state for tax, social security and labour law purposes. Moreover some studies have defined sub-categories of informal sectors in terms of basic features of units of production, average earnings, employment status (self-employment, wage employment), aims (for profit, not for profit) and social welfare coverage. The relationship with the state and its regulatory system is therefore the basis for defining and analysing informality. Usually, this interaction manifests itself as a conflict in which the 'unregulated' part of the substantive economy is essentially viewed by the state, or local authorities, as a phenomenon to be fought and reduced.

Throughout the years, economy, sociology and anthropology have focused on specific features of the informal sector, sharpening their disciplinary specialization and weakening the dialogue between different approaches to informality. Dialogue within disciplines has even proven to be difficult because different research methods highlight different features of the informal sectors, creating diverse interpretations of the phenomenon.

- Three perspectives have dominated the discussion. These include;
- The dualist schools of thought
- Structuralist schools of thought
- Legalist schools of thought

These approaches to informality are based on a series of dichotomies such as formal-informal, legal-illegal, traditional-modern and market-state. First, the dualist perspective perceives informal workers as marginal workers who - for whatever reason - failed to get a 'good' job in the formal economy and were therefore forced into 'bad' jobs of the informal economy. This informal sector essentially represents economic activities and forms of employment that were typical for pre-modern societies.

Second, within the structuralist perspective the informal sector basically represents various forms of waged employment, which are at the bottom of the employment hierarchy and thus characterized by low wages and poor working conditions. These people are excluded from state benefits and the formal labour market, constituting a segment of the workforce subordinated and functional to the currently prevalent model of capitalism. Consequently informal workers are considered to be victims of modernization and globalization processes.

Thirdly, the legalist approach sees the informal sector as a form of rebellion in reaction to the excessive and ineffective role of the state in the economy. In this view, informal workers are mainly considered as own-account workers who voluntarily and rationally decide to 'exit' the formal economy.

To go beyond the dichotomies that characterise these perspectives, more recent studies have tried to define various forms of informality along the formal-informal continuum. They have focused their attention on how the characteristics of informality are influenced by governmental regulation of labour markets, taxes and social welfare. These studies have contributed to the knowledge about informality and its functions, but theoretical and empirical attention has often remained at macro-level, leaving the analysis within the 'box' of unregulated economic activities and labour (micro-level) fairly unexplored.

Recently some researchers have tried to fill this gap. For instance in-depth studies of informal plastic waste management have revealed that local and global elements, both individually and in interaction, determine the often very conscious choice to work in the informal sector. This has allowed demystifying common assumptions about informal work being merely a result of deprivation, a lack of formal working opportunities or inadequate planning. Sometimes the informal sector provides more autonomy, economic flexibility and sustainable livelihood because it better appreciates the resourcefulness, adaptability and entrepreneurship of the urban poor.

MAJOR INFORMAL ACTORS IN MANAGEMENT OF PLASTIC WASTE MATERIALS

In addition, in order to get a source of employment for many citizens, the informal sector in plastic waste management provides an important environmental benefit to society by extending the lifetime of various products in the environment. Although most countries around the world usually do not have formal plastic management waste programs, the amount of plastic waste diverted for the purpose of re-use or recycling is very high due to the work of the informal sector. Although the informal sector is not involved in the recovery of various solid waste materials, it plays an important role in the recovery of plastic wastes. Due to growing plastic waste collection activity, there is reason to believe that today's plastic recycling rate is actually quite a bit higher.

Scavengers and Itinerant Plastic Waste Collectors

Apart from other solid waste materials, most scavengers collect plastics including PET, PP, and HDPE, and plastic bottles. The scavengers do not add value to the waste through cleansing. They do however sort the recovered materials into piles according to waste type. Once a certain sum of each type of plastic waste has been collected, it is weighed on-site and sold to middle-men who sell the goods to private plastic manufacturing businesses. For instance, in a country like Nigeria, PET and PP bottles are mostly sold to vendors of palm oil and other agricultural products that re-use the recovered PET and PP bottles and containers.

The plastic waste recovered at dumpsites is generally lower in value than plastic waste retrieved at the source by collectors. This is because most collectors divert a substantial amount of valuable materials before waste is transported to the landfill. Plastic waste materials also lose value as they are dirtied through transport.

Source:http://www.globaltimes.cn/content/1007538.shtml

Figure 7.1: Typical plastic waste scavengers.

The amount of money each scavenger earns depends upon the amount of work one is willing to put in. Some scavengers work longer hours while others choose to work shorter hours. The scavengers typically work six days a week, resting on Sunday. Rainy days affect their ability to work.

In some countries like Ghana, plastic waste scavengers together have an informal union with certain uniting principles. These include showing respect for leaders, no fighting, no stealing, and no working close to machinery. If a member violates any of these principles, punishment is awarded according to the gravity of the violation. Each scavenger contributes a small amount to the union. The money is used to cover the expenses should any scavenger fall ill or need medical.

Plastic Waste Collectors

The informal plastic waste collectors play an important role in waste management that cannot easily be taken up by the formal waste management system. Sometimes, these informal workers provide services in the absence of (effective) publically organized waste management. Due to their effectiveness, efficiency and flexibility, plastic waste collectors are integral to the waste collection activities. They are complementary to the public or private waste disposal services. Moreover, their efficiency in collecting waste - and thus secondary raw materials - is essential in times of increasingly scarce natural resources. This also holds potential to avoid the environmental hazards resulting from inadequate waste disposal, although admittedly a lot or progress can still be made with regard to recycling techniques and protective equipment.

Source: http://www.monitor.co.ug/Business/Prosper/Living-off-plastic-waste/688616-1711008-v0cb3i/index.html

Figure 7.2: Illustrating a group of women involved in collection of plastic bottles.

Plastic waste collectors' universe is very fragmented in terms of social recognition and working conditions. Research studies have shown that the informal plastic waste collectors are not passive underdogs, but that they have developed strategies to improve their economic activity. The choice to work in the informal economy might actually be an empowered choice rather than the result of sheer necessity to work informally due to a lack of formal employment opportunities, as has been demonstrated for other parts of the informal economy. However, especially the case of developing countries has shown that for some informal workers, plastic waste collecting is still a last and only resort. For them, waste collecting is a way to survive but it can also be very harmful. In these scenarios grassroots organizations provide opportunities for plastic waste collectors to improve their image and to build a 'social safety net', against for instance fluctuating prices for recyclables or customer unwillingness to pay. Collective initiatives of plastic waste collectors to change their unfavorable position can produce innovative forms of legitimization of their work and their position in urban waste management. Taking on informal waste work while still organizing themselves collectively strikes a balance between keeping their autonomy and providing better income guarantees, albeit often still precarious and incomplete.

The majority of collectors operate in the streets, going door- to-door purchasing plastic of all kinds from households and small businesses. At the end of the day, the collectors bring their goods to neighborhood collection points where the managing group of the collection point pays them for the recovered plastic waste materials.

Source:http://entecity.com/news/trivandrum-corp-to-conduct-plastic-waste-collection drive/

Figure 7.3: Illustrating piles of different plastic waste types around dumpsites.

For the collector, plastic waste is their main source of livelihood. Informal waste management is also a service provided to the community in the absence of an effectively organized formal waste management system. Informal workers see waste as an opportunity for a source of income in the gap left by the state.

Plastic waste collectors are often perceived as either victims or perpetrators. On the one hand, they are seen as victims, who are subject to social stigma and mainly considered to belong to various vulnerable groups in the society such as children, women, migrants, unemployed, disabled, and the elderly.

While the informal plastic waste collectors are recognized as a important resource in the management of various plastic waste materials in the environment, their activities are generally labour-intensive, low-tech, low-paid, often managed autonomously or in the family. They could be perceived as 'victims' of globalization, bad jobs, modernization and economic liberalization processes. On the other hand, some perceive plastic waste collectors as 'offenders'. Most local authorities and private waste management companies often have a negative attitude towards the informal plastic waste collectors and either neglect or repress them. They are considered incompatible with the increasingly standardized state and image of the current cities.

These representations as either victims or offenders are questioned (at least partially) by the collective actions promoted by non-governmental organizations (NGOs), trade unions and other community-based initiatives. However, it remains fairly unclear how plastic waste collectors are able to organize a 'collective voice' in these circumstances. The dominant stereotypes about plastic waste collectors explain neither the nature of their work practices nor their interactions with other actors in waste management. To move beyond a mere victimization of the urban poor or a one-sided romantic image of their resilience; it is important to combine a victimology perspective with a sociological analysis of the informal economy. By merging these two perspectives attention should be paid to the work practices, the attempts to organize the plastic waste collectors' actions collectively and to their relations with other social actors.

In a country like Brazil, plastic waste collectors can be divided in four types; these include

Carrinheiros standing for plastic waste collectors who work individually collecting refuse with handcarts

Carroceiros standing for waste pickers who work individually collecting garbage with rickety horse-drawn wagons;

Associations or cooperatives of plastic waste collectors

Middlemen (atravessadores): they act as informal entrepreneurs. In other words, they seem to be a crucial link, connecting informal plastic collectors to the formal industry. The sorted plastic waste changes its economic value in the hands of these middlemen. It is a source of subsistence for the informal plastic waste collectors as well as of profit in the economic exchanges among middlemen and with industry. The relationship between these middlemen and plastic waste collectors can be one of cooperation and even support, but has also shown to be fragile in face of volatile values of secondary raw materials.

The relational networks among plastic waste collectors and their suppliers ensure the continued existence of informal economic practices. The use of these forms of social capital can produce positive as well as negative effects for the informal waste collectors. In line with one of the key definitions of social capital, the fieldwork has shown forms of 'bonding social capital' such as in the interaction between plastic waste collectors and middlemen, and forms of 'bridging social capital' such as by means of the associations of waste collectors.

Carroceiros and carrinheiros are informal economic actors who operate as own-account workers and have very variable daily earnings. This means that the more plastic waste materials collectors translate to more profits.

The occupational trajectories of some plastic waste collectors show they often had previous 'formal' jobs in the manufacture or commerce sectors. Most of them usually provide a flexible service of collecting plastic waste from different place. For instance, in some countries, most of the carroceiros and carrinheiros have built a network of regular suppliers, i.e. citizens or owners of shops, restaurants, bars or offices that decided to give their sorted waste to the informal plastic waste collector instead of to the formal ones in order to reduce the risks of the recyclable plastic waste market.

In some parts of the world, the middlemen usually strengthen the dependence economic relationship between these significant informal actors by granting credit to various plastic waste collectors and provide a place for them to actually sort and store their collections. The middlemen's role is crucial because these small-scale entrepreneurs are the main buyers of collected recyclable plastic waste materials. In this production chain, middlemen thus connect the informal and the formal part of the economy. As the recyclable plastic material passes through their hands, it changes from a source of subsistence for the informal plastic waste collectors to a source of economic profit for the middlemen. Therefore, waste assumes different values throughout this process of collecting, dismantling and sorting.

The World Bank (2001) report highlighted that plastic waste collectors are crucial for urban waste management, performing collecting, dismantling and sorting actions that produce value. However, their economically weak position and social stigma considerably impede them to capitalize on this. Forms of association have therefore become increasingly viable alternatives for individual work, providing opportunities for plastic waste collectors to strengthen their position in the waste management chain.

Studies have demonstrated an 'associational diversity' that varies depending on the associations' origins, organisational resources, demands, interests and purpose. Based on the latter, plastic waste collector' associations in most part of the world fall in two broad categories:

'Inside-oriented' associations, with an original purpose of protest, basically focused on self-regulation in order to improve the allocation of internal tasks, the working conditions and to reach (albeit small) economic targets. They focus on practices aimed at reinforcing self-esteem of plastic waste collectors and monitoring and sanctioning opportunistic behaviours inside each group

'Outside-oriented' associations, which have more general goals such access to collective benefits that ensure a minimum of wellbeing despite low income.

Associations seem to grant the waste pickers a series of 'close protections', which, even though minimal and segmented, are entirely absent for informal own-account workers in most of the countries around the world. Some grassroots organizations have even gained importance beyond the mere economic. For instance, for the members of some waste collector' associations it is easier to participate in training courses organized by NGOs operating in urban areas.

CHALLENGESFACINGTHEINFORMAL SECTORS IN PLASTIC WASTE MANAGEMENT

There are several common challenges facing the informal sector in plastic waste management. They include lack of capital; lack of tools and machinery; lack of premises to operate from; poor marketing facilities and too much competition; and inadequate skills and lack of facilities to upgrade skills. With reference to informal waste collectors, health risks must be added to this list.

Access to Capital

Studies have indicated that access to capital is one of the most common obstacle limiting most informal actors in management of plastic waste materials in different places around the world. In a survey of informal production enterprises based in Nairobi, Kenya, World Bank (2010) found that more than 40% of respondents stated that their largest concern was access to credit. Other studies have similarly indicated that finance is the most common problem limiting the development of informal sector in the collection of plastic waste materials. Difficulty in obtaining access to tools and machinery are also ranked very highly.

Although capital requirements of informal sectors are lower than those of formal enterprises, banks are usually reluctant to lend to informal actors because they are usually unable to provide collateral security. Furthermore, because most informal actors in plastic and other waste management usually operate at unregistered premises, they lack title deeds required by banks. Informal actors may also lack the skills to draft the project proposals that banks need.

The lack of capital hinders the growth of informal sectors. Without capital, plastic waste collectors are unable to obtain the tools and machinery they need to add value to waste through processing. Lack of capital also makes it difficult for informal actors to rent or purchase a space wherein to store their plastic goods when the price offered by the larger manufacturers falls too low. Such a buffer against price variations can give plastic waste collectors bargaining power against the larger manufacturers. This is particularly important because the operations of plastic waste collectors are highly dependent on season. It is much easier, for example, for the plastic sachet collectors to find a large number of plastic sachets in the hot season than in the cold season since more water is consumed when temperatures are high. Similarly, plastic waste collectors are able to collect many more water bottles during holiday periods than at other times of the year since people are willing to spend more money on soda and other bottled beverages during the holiday seasons. Obtaining a storage place is of course not only a question of capital but also of access to land.

Access to Land and Accommodations

Access to space and accommodation is the second largest obstacle affecting the operations of informal sectors in management of plastic waste in different places around the world. Rapid urbanization and large income disparities in different countries has increased the pressure on land and its value and increased landlessness and homelessness. Accommodations are poor and temporary for many workers in the informal waste economy. Dumpsite scavengers typically live close to the landfills in makeshift housing. Many of the young plastic workers sleep on-site in unsafe conditions in improvised shelter.

Health Risks

Plastic waste scavenging from dumpsites and streets can pose significant occupational health risks to those involved. Actors working within the informal waste recovery system are particularly prone to respiratory problems, skin disease, eye infections, chronic backache, cuts, hallucinations, intestinal protozoa and helminthes, bites from dogs and rats, diarrhea, and jaundice. Very few of plastic waste scavengers or collectors use protective gear such as gloves, glasses, or even proper shoes.

The exposure to risk varies with the type of activity that a particular individual is involved in.

UNDERSTANDING THE FORMAL SECTORS IN PLASTIC WASTE MANAGEMENT

The formal sector in management of various plastic waste materials includes government agencies, private contractors providing plastic waste collection and disposal services, and international entities such as the World Bank, which plays an important and at times dominating role in shaping management policies in plastic and other solid waste materials in the environment. This section of the chapter discusses each of these actors in turn, as well as the challenges that the actors face.

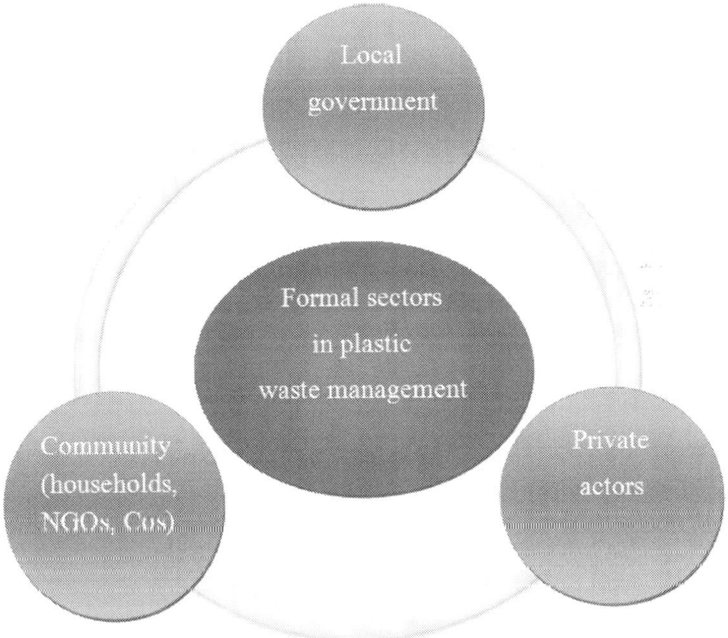

Source:https://www.researchgate.net/publication/289574916 Institutional Arrangements_for_Solid_Waste_Management_in_Cebu_City_Philippines

Figure 7.4: Illustrating key formal actors in plastic waste management.

Governmental Actors

At the national level, the government is the lead sector charged with environmental sanitation. It is responsible for the co-ordination and formulation of national environmental sanitation policy including monitoring and evaluation; the development and issuing of technical guidelines on environmental sanitation services and their management; the promulgation of national environmental sanitation legislation and bye-laws; as well as facilitating the mobilization of funds for sector plans and programs.

However, in the day-to-day management of plastic and other solid waste in most countries, the responsibility for the implementation of environmental sanitation policy at the local level is vested with the local government authority. For instance, in a country like Kenya, a policy of de-centralization and local governance has been provided in the Constitution of the Republic of Kenya, 2010, which holds that the county government is the highest political authority for a county.

Numerous other ministries and national agencies support activity on environmental sanitation. The Environmental Protection Agency (EPA) is the national regulatory agency for the protection of the environment. It is involved in the creation of environmentally-sound waste management policy as well as enforcement. Other national agencies that support environmental sanitation include the Ministry of Health; The Ministry of Education; The Ministry of the Environment, Science and Technology; The Ministry of Water Resources, Works and Housing; The Ministry of Trade and Industry; The Ministry of Tourism; and The Food and Drugs Board.

There are different challenges that different government may face when managing plastic waste materials at national level. Some of these challenges are discussed below.

Financing: According to World Bank (2001) report, the management of plastic solid waste is one of the most costly urban services to provide, absorbing approximately 1% of gross national product (GNP) and between 20 and 40% of municipal revenues in various developing countries. Cost recovery rates in plastic waste management are generally very low and subsidies typically sustain large portions of waste management. The costliest aspect of plastic waste management is transportation, followed by landfill management.

Due to the persistence of financial difficulties, most government have taken the initiative of implementing a new fee-and-performance-based policy that aims to shift the greatest share of the costs of waste management to private waste collection companies, who in turn are to shift that responsibility onto citizens. For instance, in a country like Kenya, residents are usually serviced by a door-to-door waste collection service and all residents, regardless of their level of income, are charged a fee for this collection. The principle behind this new policy, developed by consultants of the World Bank, is to "completely outsource plastic waste collection in the metropolis to private service providers", and to shift, "the responsibility of plastic waste collection and its associated revenue collection completely to the private sector". Although the national and local government are supposed to provide finances for infrastructure and equipment maintenance and replacement, major capital investments, public education and monitoring activity, the bulk of the cost of waste management (the cost of waste haulage and storage) is to be transferred to the private sector.

Regulation and law enforcement: After problems of cost recovery and lack of citizen participation, different countries face the challenge of weak law enforcement in solid waste management (especially the plastic wastes). Although the waste management bye-laws are good, some governments usually do not have enough staff members committed to enforce them. For instance, despite of various African governments taking adequate legislative steps towards addressing plastic waste management, the institutions and actors responsible for enabling and enforcing the legislation have not been effective in their work.

Nevertheless, there is certainly room for improvement in law-making as well. Most bye-laws in different countries primarily speak of the right to perform waste management services by the responsible formal actors. There is a need for a comprehensive plastic waste policy. Governments provide few mandates to private environmental service providers for dealing with plastic and other solid waste materials.

According to Ghana's 2010 revised draft of the National Environmental Policy (NEP), there are a number of highlights based on the objectives and strategic goals which are aimed at a more conscientious effort to safeguarding the environment. Policies are made as a guide to decision making to achieve rational outcomes.

Some of the key Nation Environmental Policies that have been implemented in Ghana to deal with management of plastic and other solid waste materials in environment include:

- Institutional Framework and Responsibilities
- Legislative Frameworks
- National Environmental Action Plan (NEAP)
- Financing Arrangements
- Monitoring, Evaluation and Policy Review

According to the above enumerated policies, one principal challenge remains, and that is the ineffective enforcement of the policies and laws that exist to achieve the desired goals in management of plastic and other solid waste materials.

Political interference and corruption: According to a recent study conducted by World Bank in African countries, it was found that increased privatization within the waste management sector of cities may be in part responsible for greater corruption. Although privatization has generally increased collection and disposal rates in countries, it has also resulted in greater politicization of waste management and in some cases reduced the ability of local authorities to monitor and evaluate activities. In Kenya for example, the problem seems to arise from among high-ranking appointed members of county governments.

Today there is still ample suspicion that the issuing of waste management contracts is frequently fraught with favouritism, cronyism, and nepotism in many countries. Corruption at higher levels of government hampers the ability of the responsible actors to productively fulfil their responsibilities in waste management as lower-level civil servants may be reluctant to police the activities of a particular service provider for fear of losing their job. Such corruption has impeded any meaningful decentralization and has led to a lack of responsiveness and accountability to the citizenry on the part of various agencies and their officials.

The World Bank and other Development Partners

The World Bank and numerous bi-lateral development organizations have played a large role in the formulation of crucial policies and the implementation of plastic and other solid waste projects. The World Bank has been particularly influential in urban waste management in most African and other developing countries and has worked very closely together with different governments. For instance, in Ghana, World Bank initiated a projected entitled the Urban Environmental Sanitation Project (UESP) in 1996. This project aimed to promote productivity and raise living standards in major cities, as well as to improve the capacity of Metropolitan Assemblies to manage various services involved in management of plastic and other solid wastes. The project had specific targets, one of which was the improvement of solid waste management. The total cost of the project amounted to approximately $80 million dollars and the majority of this funding came through loans provided by the International Development Association (IDA) of the World Bank. Additional funding was to be provided by the Nordic Development Fund, the Kingdom of the Netherlands, and other organizations. The repayment period was 30 years, spanning from 2006 until 2036.

According to the World Bank's project implementation completion report, project completion was "satisfactory". The project took a critical first step in the long-term process of improving local revenue collection, by beginning to build public confidence in local government responsiveness and thereby enhancing the willingness of communities and residents' to pay for urban waste management services. Today, many citizens in Ghana are quite reluctant to pay for waste services and this is of major concern to private waste collection companies.

Private Environmental Service Providers

In most countries around the world, the responsibility for plastic waste collection and disposal rest with the municipals and other private agencies. The private sector plays an increasingly large role in plastic waste management. Today, the private sector is responsible for both waste collection and waste disposal in different countries.

However, there are several challenges facing private environmental service providers in management of plastic waste materials in different places around the world.

One of these challenges is difficulty in registering households. Just as there is no bye-law requiring citizens to pay for waste services in different countries, there is also no bye-law requiring households to register with the local authorities in management of plastic wastes. Therefore, unless all households register voluntarily, the private waste service providers are force to go door-to-door registering households. This possesses some difficulties especially due to the fact that most household are usually nuclear family. Since multiple families often share one compound, family identification and registration is problematic.

Furthermore, many streets in some countries lack names and house numbers, making household identification still more difficult.

Another major challenge is lack of adequate equipment. Waste collection and disposal enterprises require high start-up capital. Trucks are expensive and require constant maintenance due to the extremely poor road conditions in most places around the world and particularly around the dumpsites.

It is also worth noting that waste collection vehicles currently being used in some countries are not designed for door-to-door collection. Many cities around the world have very poorly designed roads, many roads are unpaved, and some neighborhoods have no roads at all making vehicular access impossible. Most private service providers will find it hard to afford the necessary personnel and equipment to service these areas.

Therefore, new system of collecting and transporting plastic and other solid wastes would bring down collection rates in different countries.

Another challenge faced by private sectors is lack of technical capacity. Both the government and the private waste service companies face a dearth of qualified personnel in plastic waste management. This is both related to the fact that few educational opportunities exist for careers in waste management.

Another reason for the lack of qualified personnel in plastic waste management is that jobs related to waste receive a low status. According to the Managing Director of Waste Recycling Ghana Limited viewpoints, most of the professions related to waste management are considered "low" and that a post-secondary degree in environmental sanitation is not respected in the same way that a degree in accounting or marketing might be. According to the director, "it is quite rare to see young graduates who have the aspiration to work with anything connected to waste. In the 1960s and the 70s, it was even considered to be a stigma to work with the Ministry of Local Government since they deal directly with local politics and waste management."

Lack of citizen participation is another challenge that most private environmental service providers face in managing plastic waste in different places around the world. There are some obvious reasons for this challenge such as a lack of reliability in service provision. More significantly, however is the lack of transparency by the municipals and failure to include citizens in their decision-making processes.

One reason for the negative attitude on the part of the public towards private waste collection companies is that citizens feel that they cannot depend on these companies to provide waste services reliably. There is a common perception among many citizens that private companies want money for no collection. Due to the unreliability of the service provided by many waste service companies, citizens in low-income communities often resort to waste burning, burying, and indiscriminate dumping. When collection services falter in middle and high-income areas, citizens may pay unauthorized waste collectors to collect their waste. This is problematic because once a household has paid for waste collection; the private waste service providers usually find it difficult to collect an additional fee for the same service. Private waste collectors also lack the mechanism to off-load their waste at authorized dumpsites. They thus litter waste at container sites or dispose it indiscriminately.

Another likely reason why citizens regard most private waste collectors negatively is because they usually show little regard for the needs of citizens in their decision-making processes. Citizens are also upset with the private waste collectors because they sometimes make promises that they do not keep.

Finally, while there is a negative public attitude towards the private waste service providers, there is undoubtedly also a negative attitude on the part of the formal actors towards the public. Most people have a negative attitude towards the private waste service companies because they think they should create work for private companies to come and clean up. Rather than wanting to "create work" for the private companies, however, it is quite possible that many poor communities simply do not have the means to pay for waste services. Furthermore, they may lack sanitation infrastructure in their communities, which makes it unpleasant and unhealthy for them to store waste inside the home, waiting for collection by formal agents.

Given the private service providers failure to think carefully about how their decision-making will affect the public, it is quite easy to understand why many citizens are always reluctant to cooperate with most of them in management of plastic and other solid wastes.

AN INTEGRATED PLASTIC WASTE MANAGEMENT SYSTEM

Although considerable efforts are being made by many governments in tackling plastic waste-related problems, there are still major gaps to be filled in this area. According to UNEP (2009), the World Bank estimates that in developing countries, it is common for municipalities to spend 20 to 50 % of their available budget on solid waste management such as plastic waste materials mainly used in packaging of various goods and products. If most of the plastic waste could be diverted for material and resource recovery, then a substantial reduction in final volumes of waste could be achieved and the recovered material and resources could be utilized to generate revenue to fund other waste management activities. An integrated plastic waste management system has been pilot-tested in few locations (such as China, Pune, India, and Lesotho) and has been well received by local authorities.

It has been shown that with appropriate segregation and recycling system, significant quantities of plastic waste can be diverted from landfills and converted into resource. Similarly, the United States Environmental Protection Agency (1999) said that if a state or local government wants to plan for and implement integrated waste management system, especially when dealing with plastic wastes, it has to consider hierarchy of methods which are, reduce, recycle, and incinerate/landfill.

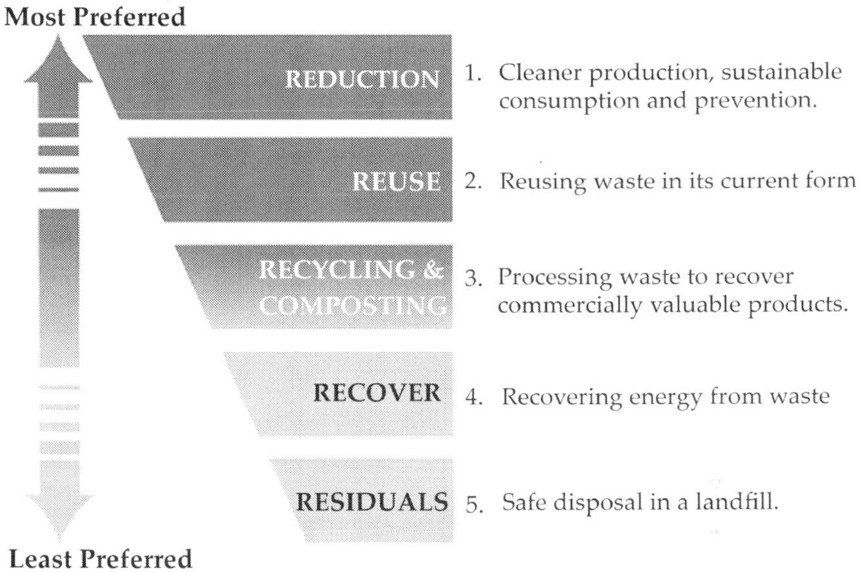

Source: http://www.frost.com/sublib/display-market-insight.do?id=186566927

Figure 7.5: An integrated plastic waste management hierarchy.

Safeguarding the environment is increasingly becoming an issue of immense concern especially in developing countries. One of the most prominent concerns resulting from not protecting the environment is high incidence of plastic waste accumulation in environment. There are several advantages that come with efficient integrated system in plastic wastes management. This may be direct benefits that can be readily valued and indirect benefits or advantages that can only be experienced. Basically, efficient plastic waste management ensures a safer environment whereby there is only little environmental pollution. Further ensuring the presence of fewer disease-causing agents or vectors and maintaining a sound ecosystem balance whereby the full benefits of ecosystem services can be harnessed.

Integrated systems in management of various plastic waste materials provides social, economic and environmental benefits to the populace, thus there is the need to recognize such systems involved, for appropriate reduction and management of various plastic waste materials discharged into the environment.

There are also several problems/challenges associated with management of various solid waste materials in different part of the world. According to World Bank (2005) report, a typical solid waste management system in a developing country displays an array of problems, including low collection coverage and irregular collection services. The many challenges confronting plastic waste management in developing countries are considered as the bed rock for their inability to rid the streets, gutters and drains of solid waste. The lack of political will of various governments to implement proper solid waste technology in developing countries can also not be over-looked. Other challenges include;

Technical Constraints

In most developing countries, there are inadequate human resources at both the national and local levels with technical expertise necessary for plastic waste management planning and operation. Many officers in charge of plastic waste management, particularly at the local level, have little or no technical background or training in engineering or management.

Financial Constraints

This is another key challenge many developing countries face when it comes to integrated plastic waste management systems. Plastic waste management is given a very low priority in developing countries. As a result, very limited funds are provided to the waste management sector by governments, and the levels of services required for protection of public health and the environment are not attained. The problem is acute at the local government level where the local taxation system is inadequately developed and, therefore, the financial basis for public services, including plastic waste management, is weak. The problem is more prevalent at the county and municipal assembly levels where the revenue generation system is inadequately developed which makes the financial basis for public services including solid waste management weak. This weak financial basis of the local governments can be supplemented by the collection of user service charges including pay as you dump system. However, users' ability to pay for the services is very limited due to high poverty level and their willingness to pay for the services which are irregular and ineffective.

Institutional Constraints

Several agencies at the national level are usually involved at least partially in plastic waste management. There are often no clear roles or functions of the various national agencies defined in relation to solid/plastic waste management and also no single agency or committee designated to coordinate their projects and activities.

Lack of coordination among the relevant agencies often results in different agencies becoming the national counterpart to different external support agencies for different solid waste management collaborative projects without being aware of what other national agencies are doing. This leads to duplication of efforts, wasting of resources and unsustainability of overall solid/plastic waste management programs. Lack of effective legislation for waste management, which is a norm in most developing countries, is partially responsible for the roles/functions of the relevant national agencies not being clearly defined and the lack of coordination among them.

8 CHALLENGES AND OPPORTUNITIES OF REUSING AND RECYCLING PLASTIC WASTE MATERIALS USED IN PACKAGING

INTRODUCTION

Due to the increase in population, lifestyle changes and social economic conditions, the generation of plastic waste has increased drastically. The major sources of plastic waste in any given scenario are manufacturing companies, IT companies, fast food restaurants, shopping malls etc. As plastic goods production is very cheap and they are readily available, their usage in day to day activities has increased over the years. But disposal of plastic waste is not as easy and cheap as its production.

Studies have revealed that today companies are facing many challenges in management of plastic waste; some of them are:

Treatment and disposal of plastic waste is the biggest challenge faced by any organization

Company face issues in deciding how to use the recycled plastic because there are variations in the quantity and quality of recycled plastic as compared with virgin plastics.

Also, the cost incurred in recycling of the plastic waste is very high as compared to other methods of waste disposal.

Different countries use different methods of disposing plastic waste materials. These include open dumping, sanitary land fill, incineration, reuse and recycling and open burning. The most common method used by countries in Europe and America such as Britain, Canada and France is sanitary land fill as opposed to open dump by countries in Africa like Uganda, Kenya and Malawi. This method of disposal has various negative impacts which include severe health risks mostly to those people living around the dump site. Burning of plastic waste materials is major contributor to urban air pollution. Burning is not only done by residents but by collectors at dump site as a method of separating out plastic materials. Health impacts can also result from direct contact with plastic waste and they include skin and respiratory disease. Randomly dumped plastic waste can pose direct hazard to children playing near dumpsite especially if the waste contains toxic or other hazardous materials.

Recycling and reusing are effective ways of minimizing problems of plastic wastes at generation point, transit and disposal, and ultimately reduces waste disposal cost. Plastic recycling can be described as "the process through which plastic waste materials previously used are collected, processed, remanufactured, and reused". On the other hand, the reuse comprises the recovery of items to be used again, in the same state but may require cleaning and refurbishing.

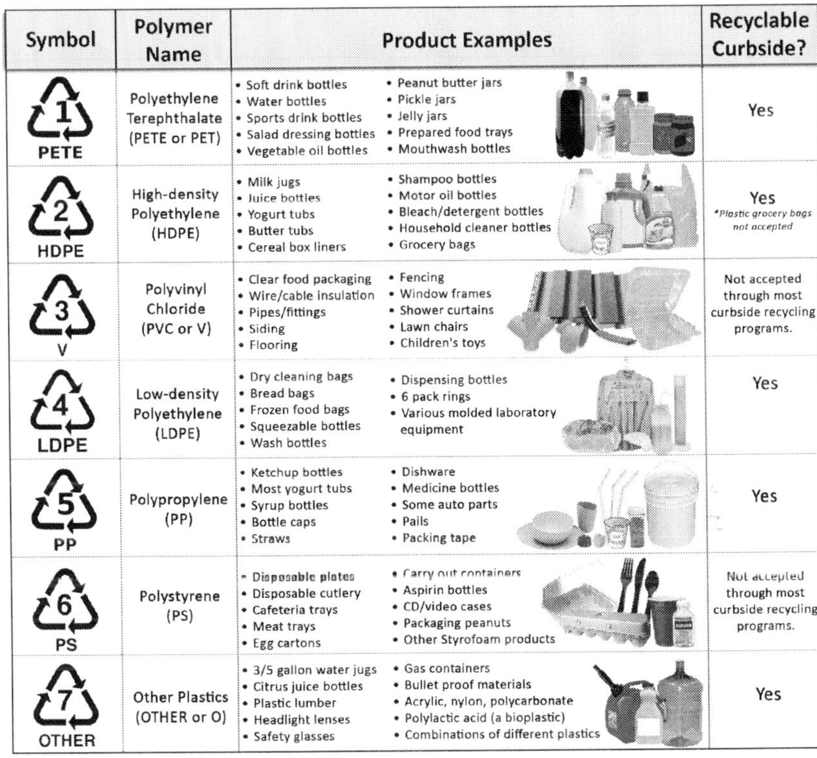

Symbol	Polymer Name	Product Examples		Recyclable Curbside?
1 PETE	Polyethylene Terephthalate (PETE or PET)	• Soft drink bottles • Water bottles • Sports drink bottles • Salad dressing bottles • Vegetable oil bottles	• Peanut butter jars • Pickle jars • Jelly jars • Prepared food trays • Mouthwash bottles	Yes
2 HDPE	High-density Polyethylene (HDPE)	• Milk jugs • Juice bottles • Yogurt tubs • Butter tubs • Cereal box liners	• Shampoo bottles • Motor oil bottles • Bleach/detergent bottles • Household cleaner bottles • Grocery bags	Yes *Plastic grocery bags not accepted
3 V	Polyvinyl Chloride (PVC or V)	• Clear food packaging • Wire/cable insulation • Pipes/fittings • Siding • Flooring	• Fencing • Window frames • Shower curtains • Lawn chairs • Children's toys	Not accepted through most curbside recycling programs.
4 LDPE	Low-density Polyethylene (LDPE)	• Dry cleaning bags • Bread bags • Frozen food bags • Squeezable bottles • Wash bottles	• Dispensing bottles • 6 pack rings • Various molded laboratory equipment	Yes
5 PP	Polypropylene (PP)	• Ketchup bottles • Most yogurt tubs • Syrup bottles • Bottle caps • Straws	• Dishware • Medicine bottles • Some auto parts • Pails • Packing tape	Yes
6 PS	Polystyrene (PS)	• Disposable plates • Disposable cutlery • Cafeteria trays • Meat trays • Egg cartons	• Carry out containers • Aspirin bottles • CD/video cases • Packaging peanuts • Other Styrofoam products	Not accepted through most curbside recycling programs.
7 OTHER	Other Plastics (OTHER or O)	• 3/5 gallon water jugs • Citrus juice bottles • Plastic lumber • Headlight lenses • Safety glasses	• Gas containers • Bullet proof materials • Acrylic, nylon, polycarbonate • Polylactic acid (a bioplastic) • Combinations of different plastics	Yes

Source:https://www.pinterest.co.uk/explore/recycle-symbol/

Figure 8.1: Illustrating some of key plastic packaging products that are recyclable.

According to Millennium Assessment Report, (2005), plastic waste recycling activities are justified on both economic and environmental grounds. This is because recycling of plastic waste can help achieve the following:

It can give people access to low-cost products and help them benefit from the technology

It can create jobs in developing countries

Reuse and recycling of plastic waste is more effective than domestic recycling.

Can help save resources, protect the environment, and contribute to sustainable development

According to UNEP (2006) report, globally 2.5 to 4 billion tonnes of plastic waste was generated in 2006 with municipal solid waste accounting 1.84 billion tonnes. As urban environmental problems worsen in developing countries, non- conventional approaches to urban pressure points like waste management will have to be adopted. The recycling of various solid wastes including plastic waste is one approach which has positive ramifications in creating informal employment and offering an environmentally sound solution to waste management problems. While there is considerable documentation on innovative community-level waste management schemes in Asian and Latin American cities, little research has been done on the importance and potential for plastic waste recycling and re-use in other developing countries such as African.

For instance, it is worth noting that less energy is usually needed to produce a new product from recycled material than from virgin material thus saving energy. For example producing one soft drink bottle from a recycled bottle, rather than raw ore saves the energy equivalent of a half-can of gasoline. The manufacture of products from recycling can also be a source of new jobs and economic development. Some studies contend that recycling produces approximately five times as many jobs as would land filling and open dumps with the same quantity of waste. Recycling programs also provide employment for people who have difficulty finding work, including those who are physically and mentally handicapped, those on public assistance, and those involved in job-skill development programs. Another benefit of recycling is that it can lead to development of an environmental ethic in the community.

ORIGIN AND THE FUTURE OF PLASTIC RECYCLING

The "reduce, reuse, recycle" that prevail in todays' society is rooted in the 1970s, a decade that was the source of both an unprecedented energy crisis and the environmental movement as it is known today. Plastic recycling began mainly through grassroots organizations that were eager to promote voluntary participation at community plastic recycling centres, and gradually grew to encompass institutionalized curbside recycling programs, state-wide plastic bottle bills, and even bans on certain plastic products. At the outset, voluntary recycling initiatives (e.g., community collection centres and curbside collection programs) were supported and funded by manufacturers, who explicitly recognized these voluntary measures as both insignificant in terms of the objective at hand (waste reduction), and as the lesser of two evils when compared to compulsory and punitive producer responsibility laws (i.e., plastic bottle bills).

According to the United States Environmental Protection Agency (EPA) (2014), both generation and recovery rates of national municipal solid waste (MSW) have increased significantly over the past 50 years. In 1960, the U.S. produced over 88 million tons of MSW, and recovered (through recycling, reuse, and refill efforts) only 6.4% of this waste. Although total MSW generation has more than doubled since 1960 to over 250 million tons in 2012, recovery rates (which now include composting efforts), capture only 34.5% of MSW. While the EPA's data suggests that recovery rates of MSW have increased steadily over the decades, other sources suggest that conventional recycling rates of beverage plastic containers peaked in 1992 at 52%, and have since decreased to a national average of 36.9% in 2010. In other words, "Americans wasted (i.e., landfilled, incinerated or littered) almost two out of every three (recyclable) beverage plastic bottles and cans sold" in 2010. It has been argued that these rates are too low, and that capturing only 20-30% of recyclable commodities will not amount to sustainable resource management. In comparison, 12 European countries recovered more than 40% of their municipal solid waste in 2012 through recycling and composting. Germany recovered the most at over 60% of its MSW, closely followed by Austria and Belgium.

To understand the implications of these rates, it has been revealed that over $22 billion worth of recyclable beverage containers alone (which does not account for other uncaptured recyclable commodities) were landfilled in the United States between 2000 and 2010. Furthermore, during this same time, "Americans landfilled, incinerated or littered enough aluminium cans to reproduce the world's entire commercial air fleet 25 times". Many scholars point out that even more perplexing is that as curbside recycling programs have increased, recycling rates have stagnated or even declined. Whereas curbside recycling collection programs serviced only 15% of the U.S. population in 1990, by 2010, programs had expanded to serve over 71% of the US population. Beverage plastic container recycling rates have being increased accordingly.

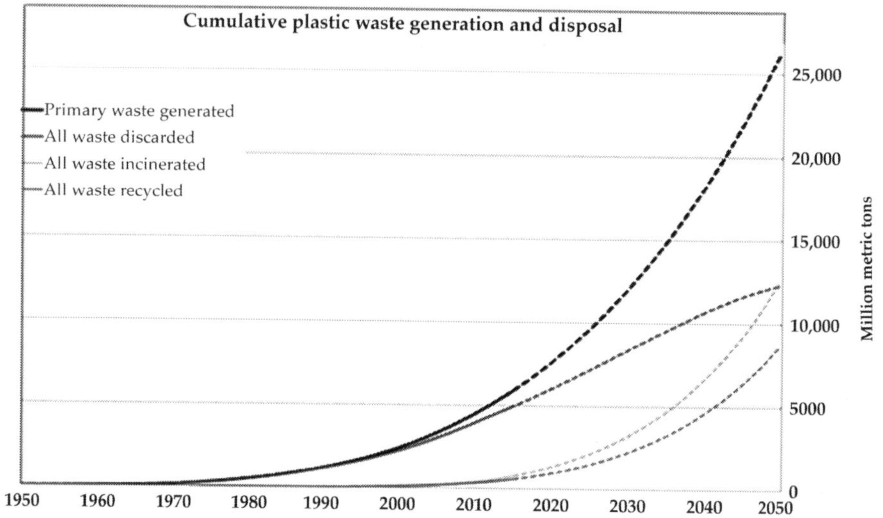

Source:http://advances.sciencemag.org/content/3/7/e1700782.full

Figure 8.2: Illustrating the trend of plastic waste production, discharged, incinerated, and recycles.

The vision for the future of plastic waste management's transformation to thorough materials management, while useful in terms of understanding the fractures and opportunities in the current paradigm, is only a vision. The reality, on the other hand, involves the current state of plastic recycling in which two out of every three beverage plastic containers are wasted instead of recovered. Thus, many studies have focused on examining and offering methods of increasing recycling rates in the United States and abroad. Some studies suggest that many individuals are found to self-report pro-recycling behaviors, while not practicing pro- recycling behaviors. Others suggest that this disconnect requires that society increase the social stigma of not recycling.

Overall, many studies focus on how to target a variety of pro-environmental behavioral changes in consumers, and ultimately attribute the power to the consumer to shift manufacturing activities. Others would argue these premises are unfounded and contribute to the prevailing hegemony of covert manufacturing power in the broken recycling paradigm. Still yet, some studies focus on the potential repurposing of commodities tough to recycle, such as converting plastic into fuel.

Nevertheless, a consensus exists among the various studies that specifically analyze how to increase recycling rates: signage near recycling bins that delineate which materials should be landfilled and which should be recycled is minimally effective, at best. Modeling or added social pressure via covert researchers who "model" the desired behavior in the public setting being studied, can be very effective at increasing plastic waste materials recycling rates. Modeling contributes to the norming process many scholars discussed by creating a social system that fosters pro-recycling behaviors. Similarly, interventions such as personal feedback for households in curbside collection programs, while costly and resource-intensive, can help increase recycling tonnage.

While useful, these studies and suggestions involve many limitations. Some studies boasted increasing recycling rates, but failed to conduct audits of the waste stream to ensure that the changes were indeed due to increases in recyclable commodities as opposed to increases in contamination from non- recyclable commodities. Furthermore, many highlighted the remaining need for increased education regarding recycling, and some even pointed to better signage at bins, neglecting studies that demonstrated negligible impacts of signage on recycling rates.

Educational efforts to increase pro-recycling behaviours, especially carried out through means other than signage, require money, time, and personnel that not every municipality, business, or building has. For instance, a Minneapolis-based non-profit, Recycle Across America (RAA) has started a movement to standardize labeling for plastic waste and recycling bins across the country. While impressive, these standardized labels fail to reduce confusion regarding how curbside collection programs differ by state, municipality, hauler, or even building, because they attempt merely to offer a standardized way of conceptualizing and visualizing different waste streams, rather than standardizing what belongs to which waste stream nation-wide.

SUSTAINABILITY IN PLASTIC RECYCLING PROCESS

Sustainability supports the integration of economic, environmental and social concerns into a unified framework for responsible decision-making and marks a fundamental shift in policy towards a long-term agenda. To be successfully implemented, governance that is transparent and held accountable for the results of municipal programs must support sustainability. Different countries have taken efforts to support the quality of life for all citizens in respective countries while cultivating long-term growth. Unfortunately, the goals to a sustainable plastic waste recycling process have consistently underperformed without obvious steps to remedy the problem. For fact that needs to be understood is that the generation of plastic waste in most countries usually go undocumented and plans do not address the root of the problem.

While some aspects of sustainability are well understood, environmental resource management and recycling participation has fallen short. Therefore, programs that support responsible long-term environmental decisions should be adopted in order to have sustainability in plastic waste reduction in the society.

Recycling and Sustainability

Plastic waste recycling helps in promoting sustainable consumption and production practices, resulting in valuable economic and environmental benefits. To enhance sustainability of plastic recycling, plastic separation at the source point is uniquely suited, especially in an urban context. This is because the logistics can be integrated into daily routines, and it represents one of the few ways for people to steward responsible plastic consumption within an urban setting. Studies have shown that early childhood education in sustainability must be rooted in local context for it to be meaningful and have impact on behaviors. Therefore, managed school recycling programs can be used to reinforce plastic recycling habits and integrate sustainable education into daily activity. In addition, plastic recycling education helps connect resources to consumption, which could influence future source reduction.

Implementing a plastic recycling program not only shapes current plastic waste management, it also has the potential to influence and improve recycling performance for many years to come. Educating students about recycling reflects a solid waste management strategy to impact long-term behavior.

In an urban environment, plastic recycling process is a low cost; low-technology means to apply values and principals vital to the future of our planet. Awareness creation or educating people on plastic recycling sustainability increases the capacity to address environmental concerns. It can acts as a significant way of empowering young people to become stewards of their environment. Recycling is a natural fit with existing education agendas both ethically and practically. According to UNESCO (1997) report, education can be used as a humanity best hope and most effective means in the quest to achieve sustainable development.

Environmental Education

Educating people on sustainable choices through a plastic waste recycling program empowers them to address pressing environmental issues. In a country like Thailand, government and non-government organizations' projects, like 'Waste Minimization,' 'Thinking Over and Saving Resources,' and 'Community in Your Hands,' aim to increase household education related to reducing, reusing, and recycling plastic and other solid waste materials. With action-based initiatives, plastic waste recycling programs support both general education and environmentally driven agendas. A well-developed environmental education program has the potential to develop communications, teamwork and problem-solving skills and is therefore important towards developing good citizenship. Education on environmental concerns that foster solutions through collaboration has the potential to facilitate active community engagement. To achieve the greatest impact, local context should determine how sustainability is incorporated into educational programs, both at community level and various learning institutions. In an urban setting, plastic waste recycling can augment environmental education and awareness while developing a deeper understanding of sustainability through a collaborative effort.

Sustainability as Habit

For sustainability in plastic waste recycling process to move outside of a conceptual framework, it must become integrated into habits that guide daily actions. Plastic waste recycling programs in learning institutions can actually be used in providing learners with the opportunity to implement sustainable principals within daily routines and empower young people with knowledge and experience to actively contribute to a sustainable future. Fostering responsible citizens, who understand behavioral choices as the basis for sustainable implementation, provides a vital method to invest in the future. Plastic waste recycling provides a path to actively take part in the creation of a sustainable society, through concrete choices grounded in familiar activities.

Quantification of Plastic Waste

Quantification of various plastic waste materials is a necessary step towards have a sustainable process in managing plastic disposal and developing effective policy. There are several descriptors that can actually be used to clarify areas of plastic waste materials quantification. These include quantifying plastic waste generation, contamination, participation and diversion.

Plastic waste generation is the act of creating material that is no longer desired and therefore does not have value in its current form. These materials are therefore gathered for disposal. Plastic waste generation rate refers to the amount of materials produced by a group of individuals over a given period of time. The rate can be calculated for the amount diverted, disposed, or produced without necessarily indicating end processing. Plastic waste generation is expressed as mass per population producing waste per time interval.

Contamination on the other hand refers to the amount of material improperly disposed of based on local waste management regulations. Contamination of potentially recyclable plastic material within the waste stream for general disposal increases the amount of processing due to extra handling. High levels of contamination may prevent this material from being properly processed, which increases the adverse environmental impact of disposal and the amount of labor and cost required to prepare this material for processing.

Participation in a plastic recycling program refers to the percentage of a certain group that correctly source separates various plastic recyclable materials from non-recyclable materials as required by the next step in the waste management process. The rate of participation can actually be used to assess the success of a recycling program as a whole or of individual components. In the case of a learning institution, participation rates can be calculated for students, custodial staff, classrooms or offices.

Participation quantifies the value of a recycling program, signifying social engagement and potentially assigns responsibility for non- compliance.

Diversion rates quantify the results of overall waste management practices as a percentage of waste that is kept from landfilling and indicates the effectiveness of a reuse and recycling program.

Lastly, the rate of diversion in a recycling program can be quantified based on the total amount plastic waste materials that would have gone to landfill or as a comparison with what has been disposed of previously. The data needed for these statistics can be obtained through a plastic waste audit that specifically examines waste disposal patterns. Diversion rates however can be deceiving if more information is not provided. For instance, a country like Ghana diverts nearly half of its plastic and other solid waste materials from landfills. It is therefore important to indicate what plastic waste materials are diverted and the capture rate for these materials that should be source separated from the waste stream. Recycling rates are also useful benchmarks for quantifying the progress made when implementing a plastic waste recycling program. The diverted recyclables may also translate to solid refuse reduction, which has specific environmental and financial implication based on location.

Plastic waste management: The term recycling is used to describe the act of separating a material from the waste stream headed for landfill or other processing so that this material may be processed and recomposed into a useful new material to create a product. Materials diverted from the landfill are recovered to have another function, which may or may not be similar to its original function and appearance. Recycling programs increase the amount of materials available for reuse and therefore potentially avert the consumption and extraction of virgin materials.

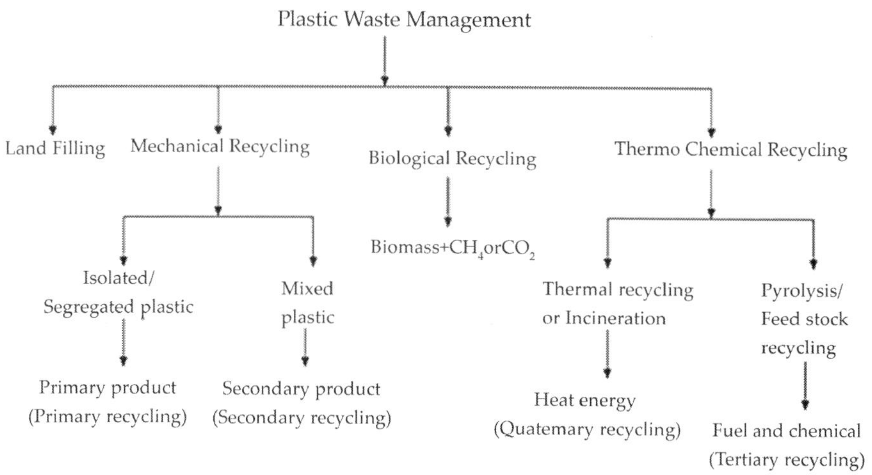

Source: https://www.researchgate.net/figure/227421226_fig2_Fig-2-Different-routes-for-plastic-waste-management

Figure 8.3: Different routes for plastic waste management.

Recycling plastic materials and reducing waste has the potential to reduce waste management costs and to increase the lifespan of existing landfills. An effective management plan for plastic waste materials includes the highest possible diversion rate away from landfills with minimal transportation and environmental impact. The system of getting plastic waste on a path to diversion is vital to increasing participation. An efficient plastic waste management system must pinpoint the weak links and strengthen these connections. Recycling plastic waste recovers materials to be used again and averts pollution. As landfill space has become more precious, recycling has become more important. Proper waste management requires an understanding of the composition of the waste stream and how it was generated. Plastic waste management strategies therefore must be based on composition, which reflects where it was generated, the local market, and management options.

KEY ELEMENTS OF PLASTIC WASTE RECYCLING PROGRAMS

Motivation

One of the elements of successful plastic waste recycling programs comes through the motivation, which has a direct impact on the ambition and the results of the program. It is worth noting that a recycling program that shifts the financial benefits of recycling process is likely to experience success through the higher motivation and participation of those involved. A powerful financial incentive provides the catalyst to begin and maintain a plastic recycling program. The quantifiable results are a testament to the strength of finances to motivate behavior. For the municipals that are responsible for the cost of plastic waste management, the money saved from recycling becomes representative of success. As the disposal of plastic waste is quantitatively assessed in the disposal process, areas of a country that experience financial benefits due to recycling of plastic wastes are likely to also report on the plastic waste tonnage.

There are some environmentally sensitive areas of the country that are likely to also have more ambitious recycling agendas as part of a sustainable vision. One instance, some residents in Nairobi area in Kenya have being pursuing an ambitious zero-plastic waste agenda. Most of these residents were motivated by the location of Nairobi city, which is in an environmentally sensitive wetland, the residents wanted to responsibly manage plastic waste materials through recycling and reusing. This program has resulted to about 85% reduction in plastic waste produced as related to the broad vision of the program, which addresses the many different solid waste flows.

Higher rates of participation in a recycling program are usually experienced when the recycling program is integrated into a sustainable education agenda to which recycling becomes a tool to apply the belief system that has been cultivated. This means that integration of plastic waste recycling programs into lesion taught in various learning institutions is likely to support the larger impact a recycling program can have to improve plastic waste management in the larger community and in the future.

Administrative

Administrative elements must also be in place for a recycling program to succeed. Strong leadership, administrative support, and clear policy must be in place to support it.Poor internal communications and the misallocation of bins to places that are not visible or easily used may actually result into low participation and low awareness of the recycling program. However, the logistics portion is not necessary in all plastic recycling programs. This is because some recycling programs are usually revamps of existing programs that may underperform for other reasons. There are certain barriers that are likely to be experienced in investing in a sustainable plastic recycling project. These barriers include lack of money, time, and commitment. For the administrative phase to be successful, the commitment of the administrative staff must already be in place.

Plastic waste reduction at the source point should play a part in the planning of a recycling program as consumption factors also affect the waste stream. Consumption varies based on convenience, the attitude of the users, perceptions of more pressing issues, concern for the environment, as well as operational challenges that limit implementation process of the recycling program. Planning for behavior changes must also consider the consumption factors that drive plastic waste disposal. For instance, plastic waste reduction should be addressed as part of the larger vision for sustainable management, of which recycling is a part.

Creating multiple opportunities for participation and different levels of involvement helps to ensure the continuity of a sustainable initiative. Implementing a plastic recycling program may require multiple participants based on the multiple roles that are required to successfully develop a recycling program.

The first role requires researching about the impacts of plastic waste materials and making decisions considering environmental impact

Second, those individuals involved should be able to either supervise or perform implementation.

Finally, a coordinator must monitor implementation and make changes as required

A plastic recycling program is likely to fail due to lack of engaging all those responsible stakeholders, especially when it comes to determining how the program works. Lack of administrative support may undermine efficacy of a recycling program.When initiating a recycling program, it is advisable to come up with a recycling team. A recycling team is one of the most important determinants of success and longevity of a plastic waste recycling program. Although important, the designation of a Recycling Coordinator is not enough to build a successful recycling program.

In most cases, plastic waste audits are rarely conducted when not recommended by those guiding the development of a recycling program. Therefore, lack of characterization of plastic waste suggests the need to document waste composition to generate the necessary data to formulate better management strategies. The waste management literature consistently reinforced the importance of waste audits in the development of a new program.

Implementation

To implement a plastic recycling plan, changing attitudes and behavior is fundamental. Plastic waste recycling plans not only change attitudes so that those involved are open to breaking habits, they also smooth and facilitate this transition. The most successful plans remove situational barriers, provide information on how to properly recycle plastic waste materials, and set goals that increase awareness and motivate participants. There are other factors that are likely to influence whether people choose to participate in a plastic waste recycling program. These factors include;

- Level of awareness
- Parental behaviors and influence
- Community interconnectivity
- Concern for the environment
- Perception of ability to make a difference
- Effort required
- Perceived effort

A recycling program must be deployed to change attitudes first and behavior second with an awareness of the impediments to implementation.

The success of a recycling program depends not only on the participation and diversion rate, but also on the efficiency of a program, which will be shaped by the time commitment of those involved. This can be expressed as the benefit to cost ratio, which in this case is the cost of processing per ton. Effectiveness will be based on the time it takes to break habits and establish new routines as well as time to coordinate new logistics. For a recycling plan to be successfully implemented it must be time efficient for those involved especially the custodial staff. Material costs therefore are a subset that improve effectiveness and reduce the time required for implementation.

Maintenance

The maintenance of a plastic waste recycling program is vital to its long-term success. Advertising helps to ensure the long-term interest and commitment of those involved. In programs that have been successfully implemented, it is often expressed through statistics representative of staff experience with minimal published introspection or quantitative analysis. Visible results, which offer public relation benefits, are often touted as well. The municipal or any other institution responsible for plastic waste recycling program may be hesitant to advertise recycling rates when diversion rates are dismally low. Except in cases where a recycling program is formed and monitored, a legal impetus is usually required to document progress in the form of a waste audit.

RECYCLING-REUSING PLASTIC BEVERAGES BOTTLES

Several proposals and schemes have been made to reduce the disposal of waste in landfills. Perhaps the most significant system for waste reduction is the one that promotes recycling and reuse of discarded items. Plastic beverage bottles are one of the largest components of waste discarded by human beings. According to a survey conducted in 2011 by Container Recycling Institute revealed that "Americans throw away 200 billion beverage plastic each year. Beverage plastic containers make up about 15 percent of all packaging waste in the US, and in the year 2000, only 31 percent were recycled" This suggests that 69% were disposed into the environment, which is alarming. Customers' consumption behaviour and disposal habits have not changed significantly enough to reduce disposal. Therefore, the assumption can be made that plastic bottle waste worldwide is exceedingly high.

Currently, there is an increasing focus on the importance of recycling and reuse in an effort to save the environment from the harmful substances that result from plastic waste disposal activities in landfill locations, which are becoming less available with time. Many cities have created a new system for waste collection where recyclables go in one bin, non-recyclables in another and food scraps go in a third. Some of those recyclable items are plastic bottles. When disposed of in landfills they take hundreds of years to degrade rendering such lands unusable for those many years. In addition, as these bottles degrade they admit harmful chemicals into the environment.

In other words, most plastic beverage bottles are produced from a material known as "Polyethylene terephthalate" (PET) which contains a harmful chemical known as "Bisphenol A" (BPA). As plastic bottles decompose/break-down in a landfill BPA is admitted into the surroundings of the landfill and contaminates the environmental water table.

Additionally, there has been a growing concern for the various companies that manufacture soft drinks to be responsible for their processing activities and to reduce the amount of the plastic beverage bottle that go as waste into the environment. For instance, large soft drink companies such as Coca Cola Company, PepsiCo Incorporated and Dr. Pepper/Seven-Up Incorporated have embraced the use of plastic bottles to contain their product that is sold on the market. In 2002, "Pepsi began using ten percent recycled content in its bottles, which Coca Cola was already doing. Since then, Coke has promised to up that recycled content to 25 percent in the US – Coca Cola already used bottles with 25 percent recycled PET in other countries".

In recent years, several studies that deal with recycling and reuse have been published. Also there have been several other studies that suggest the use of alternative materials such as biodegradable plastics. To effectively achieve the needed goals and results in recycling plastic waste bottles and other plastic materials used in packaging industry, there are a few components that should be put in place. Some of these components are discussed below.

Community Participation

According to World Bank (2006) report, different people according to their cultural context define community participation in communal activities differently. Community participation means involving people; men and women in the development process as active participants and not as passive recipients at all levels. Studies show that if well planned and executed, plastic wastes can profitably be re-cycled. In Latin America, for example, cooperatives and NGOs are actively engaged in the collection and separation of wastes in small scale recycling enterprises. In Brazil and Argentina CBOs have emerged with a component of refuse collection, separation and recycling. Development which requires participation can no longer be left to chance or to a few groups of individuals. This means that any effective and meaningful participation in development must involve different players.

Source:http://bostonmountain.org/2016/03/25/green-bench-challenge-gets-students-excited-about-recycling/

Figure 8.4: Illustrating community involvement in plastic waste materials re-cycling program.

Moreover, literature suggests that voluntary or other forms of contributions are indeed fundamental to any development activity success. This means that voluntary contributions of labour may cover a wide range of practices, from forced labour with a legally enforceable penalty for default, to work in a group voluntarily joined without pressure.

Awareness Creation

Community participation and awareness creation go together. Awareness means exposure of an individual to an idea. For instance, a community may know the nature of a plastic recycling project but may not know the details. Awareness makes one develop interest that is he/she becomes motivated to find more information about the new idea. It has been noted that any development program could be effective only when people are aware about it and the benefits that will accrue to them as a result of implementing it.

Recycling Team Formation

Studies show that if well planned and executed, a plastic waste recycling project can be implemented and managed successfully. In Latin America, for example, cooperatives and NGOs are actively engaged in the collection and separation of different solid wastes in small scale recycling enterprises. In Brazil and Argentina, Community Based Organizations (CBOs) have initiated different plastic recycling projects involved in collection, separation and recycling/reusing plastic waste materials both at community and national levels. Many recycling teams from different parts of the world have seen the advantage of integrating recycling projects into various countries' strategic plans. Therefore, formation of recycling teams will not only help in reducing plastic waste materials in the environment, but will also ease the work of various responsible authorities mandated to manage and control waste discharge into the environment. Plastic recycling teams also offer the municipalities an alternative waste management tool and creation of micro enterprises. Therefore, the idea of creating a plastic waste recycling team is to mobilize communities to collect and make use of various plastic beverage bottles that are disposed as waste.

FACTORS INFLUENCING SUSTAINABLE PLASTIC WASTE RECYCLING

According to UNEP (2001) report, "plastic waste is an inevitable by-product emanating from various human activities. In the past, this was not a major problem because almost everything was reused or recycled and whatever remained was taken care of by nature." There are a number of factors that influence recycling processes of plastic waste materials in a community or a country. These factors are outlined below as follows.

Government Finance

The various operations involved in plastic waste recycling process, allocation of budget by the national and local government, funds reliability and stability are some of the key government financial factors that influence plastic waste recycling processes especially in developing countries. In China for instance, the government allocated 60 billion Yuan (approximately 7.4 billion USD) to environmental infrastructure such as environmentally sound waste treatment facilities in 1998. Aside from China, another case study demonstrating an incentive to recycling through government finances is Jamaica, where funds from tax on imported plastics is apportioned to solid waste management as of 2004. Brazil has found that curbside collection of recyclable plastic materials is more costly than collecting mixed waste, but has recognized that the small additional amount paid to recycle the material is less than the combined monetary expense of landfill space and operation and environmental cost of less sustainable solid waste management.

Plastic Waste Collection and Segregation

Collection and segregation of solid waste is another important factor that influences recovery of materials. Collection and segregation of plastic waste materials in developing countries is usually done by scavengers, municipalities, or private contractors and can be considered formal or informal, depending on the degree of organization in any of the three sectors. Plastic waste collection and segregation can be viewed as an incentive or barrier depending upon technological and human resources available in a given case study. Countries that employ a more technological approach of material recovery (that is, waste segregation with technical machinery); efficient plastic waste collection is seen as an incentive because all waste is eventually directed to a processing facility. On the other hand, countries that utilize labor-intensive means of recycling (such as manual separation) may not necessarily require gathering waste into one location due to the wide dispersion of scavengers throughout a community.

Since the accumulation of waste in dumpsites near residential areas poses a major health concern, higher waste collection efficiencies should be considered as an incentive to recycling rather than a barrier. This assumption still affords scavengers the opportunity to collect material of value in the waste stream from a common area, which may promote organization and formalization of this vital player in sustainable plastic waste management, especially in the third world.

On the other hand, segregation of various plastic and other solid waste materials is influenced by the extent of technological advancement of a country. For instance, if a facility or series of facilities is capable of separating organic from inorganic materials, recyclable from inert materials, cleansing recyclables, and then processing the recyclables into 'raw' material for sale, then mixed waste collection is sufficient. Since developing countries usually lack such technological advances, segregation of waste at the source by divided or selective collection or segregation of waste at an intermediary step by scavengers is necessary.

Household Awareness and Economics

According to a report by UNEP (2001) it is worth noting that regular people are likely to take part in an environmental activity that is meant to help the society if they are well informed about the benefits of the activity. Having crucial information and understanding the relationship between the behaviours of human being, how they handle various solid waste materials and environment/health/sanitation benefits are some of the key aspects of household education that are likely to influence plastic waste recycling activities. There are several countries that have embraced the importance of household education in enhancing solid waste recycling. Some of these countries include Sri Lanka, Mongolia, Indonesia, Thailand, and Mexico among others. In Thailand for instance, government and non-government organizations' projects, like 'Waste Minimization,' 'Thinking Over and Saving Resources,' and 'Community in Your Hands,' aim to increase household education related to reducing, reusing, and recycling plastic s and other solid wastes. Additionally, administrators in Jamaica overwhelmed by the littering behavior of citizens look to household education campaigns to alleviate their solid waste materials problems.

Apart from household education, studies have revealed that household economics play an important role in recycling of various solid waste materials. In other words, the level of household income influences the behavior of handling various solid waste materials (such as reusing, recycling, and illegal dumping). For instance, in China, the implementation of a fee system of 3 Yuan Tax/family/month for treatment and disposal of solid waste had positive effects on the awareness of waste problems, which in turn encourages participation in other activities to reduce waste such as recycling. Other case studies include low-income households in Botswana, Guyana, and Iran, the idea of 'waste' is uncommon because every material commonly present in form of solid waste is seen as valuable. Also, the role of scavengers as informal recyclers continues to be a result of citizens in need of income. In some cases, as in the case of India, increasing incomes can act as incentives to recycling also due to changing consumption patterns that result in a higher percentage of recyclables solid plastic wastes materials.

Administration of Solid Waste Management

The existence and the effectiveness of either public or private management of waste through collection, recovery, and disposal is another very significant factor that influences plastic and other solid waste recycling activities. A barrier to recycling within administration of solid waste materials is the lack of collaboration among the various agents collecting, treating, disposing, financing, and regulating wastes. Such barriers were evidenced in Indonesia, Iran, Mexico, Nepal, and Vietnam due to the unstable/undeveloped political institutions and limited/unreliable financial resources. In some places, private administration of solid waste materials has really help in alleviating the work of recycling plastic waste materials. This therefore means that private administration is used as an incentive to waste materials recycling, especially in developing third countries. Private administration alleviates pressure from the public administrators by making recycling profitable, modernizing recycling process (such as being able to afford machinery), and reducing the number of responsibilities.

Marketing the Local Recycled Materials

Marketing of the recycled waste materials is another factor that influences recycling activities of solid waste materials in developing countries. For instance, in Jamaica, a small-island developing state, management of various solid waste materials is comprised of 40% recyclables, but no viable markets exist due to the overall low quantity of recoverable materials. Conversely, Sri Lanka, also an island nation, experiences substantial export demand for plastic waste products.

According to World Bank (2003) report, in Thailand, it is estimated that the market value of approximately 4.5 million tons of recyclables would near 16 billion THB (Thai Baht) per year (approximately 390 million USD). Likewise is true in Iran, where "if an official system was put into place for plastic waste recycling, one could earn USD 44 million per year".

Technology and Human Resources

The effectiveness and the availability of technology and human resource are key factors that warrant the success of any plastic and other solid waste recycling program. In other word, both technology and availability of human resource are key influencers of a sustainable solid waste recycling program/project. In many cases, especially in various developing countries, sufficient funds act as a hindrance to the implementation of technology in a recycling project. Even when a capital investment can be made for a piece of technical equipment, many times laborers are unskilled or funds are undependable to operate and maintain the equipment properly. Both instances pose health and safety hazards, and both are seen as barriers to recycling of plastic waste materials.

Different governments advocate for the use of manual labor in place of technical machinery in developing countries, such as China, Guyana, Jamaica, Lebanon, and Turkey, whereas others simply advocate for better technology, namely India, Indonesia, and the Philippines. Countries such as Laos and Mexico advocate for the use of collection vehicles and manual solid waste materials collection, respectively. In any case, with or without technological inputs, institutions must be in place to safeguard the health of the technicians and manual laborers.

Availability of Land

Availability of land is usually seen as an incentive to recycling. It acts as an influence to management of solid waste materials. Poor terrain, widespread development (population density), and extensive private ownership are usually seen as hindrances to landfill management of waste. Such constraints force municipalities to consider other options of managing solid waste materials such as recycling and incineration with or without energy recovery.

In Lebanon for instance, the cost of land is very high. This is because most of it is privately owned, which is prohibitive to the administrators obtaining it for landfills. Additionally, in Brazil, rugged terrain causes most Brazilians to live within 100 kilometers of the sea. "This coast-hugging pattern of human settlement has exerted an increasing pressure on available space to bury much of the 90,000 tons of solid waste materials that Brazilians generate every day". Sri Lanka and Mauritius on the other hand are two small-island developing nations that face issues of high importation of goods and tourism resulting in increasing amounts of solid waste materials without a management solution in place.

9 SUSTAINABLE PLASTIC WASTE MANAGEMENT THROUGH ENERGY RECOVERY TECHNOLOGY

INTRODUCTION

Sustainability has become an integral part of building corporate strategies regarding energy use, supply chain, and waste management. Considering sustainability is advantageous for corporations by addressing the concerns of a growing number of consumers, benefiting economically, and managing risk. According to United Nation (UN) "sustainability means to meet the present needs without compromising the ability of future generations to meet their own needs," this has been a foundation upon which to build corporate strategy. A corporate look at sustainability takes a multi- dimensional approach, also referred to as the triple bottom line (TBL), encompassing economic, environmental, and social domains. Studies have revealed that "triple bottom line (TBL) and its core value of sustainability have become compelling in the business world due to accumulating anecdotal evidence of greater long-term profitability. For instance, reducing plastic waste from packaging through energy recovery technology can reduce costs. Among the key organizations that have used this approach in their plastic waste minimization strategy include General Electric, Unilever, Proctor and Gamble, 3M and Cascade Engineering.

On the other hand, energy recovery from waste is achieved by a range of technologies including landfill gas recovery, incineration, and gasification. According to U.S. Environmental Protection Agency, (2012) report, electricity production from captured landfill gas is the most widely implemented form of energy recovery in the United States. Additionally, there are approximately 75 facilities producing electricity from captured gas in California with a total capacity of 299 MW. Typically landfill gas is composed of 57% methane, 42% carbon dioxide, 0.5% nitrogen, 0.2% hydrogen, and 0.2% oxygen, along with trace quantities of other compounds. Some studies report that landfill gas collection is a better energy recovery strategy than incineration because materials like plastics and fibrous organics continues to provide carbon sequestration as they are effectively entombed in the landfill.

Anaerobic digestion (AD) technologies provide both material and energy recovery in the form of compost and concentrated digester gas, which can be used to generate heat and electricity. These systems accelerate the biological decomposition process using heat and microbes to reduce the time needed produce compost and digester gas. Digester gas contains higher concentrations of methane than landfill gas and is a viable alternative for natural gas. Aerobic digesters are best suited to process wet biogenic wastes and are commonly deployed at wastewater treatment plants and dairies. Anaerobic digestion is not as effective at processing woody biomass, which is more suited for other forms of composting or energy recovery.

Incineration provides energy recovery by capturing thermal energy from combustible materials in the waste stream, such as plastic waste materials. Other by-products include water vapour, char, fly ash, carbon dioxide, and hazardous gases such as nitrogen oxides, sulphur dioxide, particulate matter, carbon monoxide, acid gasses, lead, cadmium, mercury, dioxin, and furans. Exhaust scrubbing technologies have improved over the years, dramatically cutting the pollution levels and emissions from incinerators. Even with these improvements, incinerators are still viewed by the public to be incredibly harmful to the environment. The majority of the incineration facilities in operation are on the east coast where there is a higher population density and less space available for landfill development.

Gasification is alternative form of energy recovery that uses a non-combustion thermal process to capture chemical energy from waste. According to U.S. DOE National Energy Technology Laboratory, (2010), in the last twenty years, gasification technologies like pyrolysis, conventional gasification, and plasma arc assisted gasification have gained more credibility on the world market. Gasification technologies capture chemical energy by heating feedstock waste materials in a low oxygen environment to produce a gas mixture of predominantly carbon monoxide and hydrogen, referred to as synthetic gas or syngas. The syngas, once cooled and cleaned, is a versatile energy source.Syngas can be combusted to produce heat and electricity in a Brayton cycle, Rankine cycle, or internal combustion engine. Syngas can also be converted to liquid fuels, and research is underway for using syngas in fuel cells. Depending on the gasification process, the other marketable by-products are fly ash and char, or slag, comprising one tenth of the original volume. The high temperatures of gasification reduce the leach-ability of these by-products and can therefore be repurposed as an additive in a cement or construction aggregate.

EVALUATION OF ENERGY RECOVERY PROCESSES

According to a widely-circulated waste management hierarchy, the most preferred option for managing plastic waste is to reduce and reuse. If those strategies cannot be applied, recycling followed by energy recovery are the preferred options. Plastics waste stream of diversion focus on PP and PVC containers with feta residue falls into the energy recovery portion of the waste hierarchy because food residue and material type prevent recycling. Researching energy recovery possibilities for plastic waste is the emphasis of this chapter. Different scholars have discussed practical examples of plastic waste energy recovery technologies. For example, there two widely used techniques on how energy can be recovered through the use of plastic waste materials. These are;

Converting plastic waste materials to crude bio-oil through pyrolysis

Using non-recyclable plastic waste material to create SpecFuel (a pellet fuel used as a coal-substitute)

The technology that is used in recovering energy from plastic waste materials is known as 'anaerobic thermal reclamation,' which uses an oxygen-free chamber and heat to process the plastics. Contaminants such as dirt, fibre and food are not problematic, so there's no need to clean the plastics before processing. The plastics travel by truck to the plastics recovery facility, where technicians can turn plastic into crude oil in less than five hours. The final step is loading the oil into tanker trucks and hauling it to refineries". SpecFUEL is another option of recovering energy from plastic waste materials, where post-consumer plastic waste materials are transformed into a high-BTU pellet to be burned in place of coal, petroleum coke or biomass. Because plastic waste energy recovery process does not require cleaning of residues from plastic before processing, these technologies seem worthy of further research. A scientific review of pyrolysis and gasification supports the technical feasibility and industrial value of the processes' products; however, the lack of commercial availability near the recycling facility and the need for increasing efficiency through pre-processing plastic waste may lead to focus on refuse-derived fuel pellets. United States Environmental Protection Agency (2013) gave a broad definition of refuse-derived fuel (RDF), "a nearly unlimited broad range of solid, liquid and gaseous waste materials from household, commerce, forestry, agriculture and industry, which have a certain calorific value, may be applied as 'waste fuel' or RDF in Waste to Energy or co-incineration plants after having undergone different levels of prior processing." plastic waste palletisation can be helpful in terms of reducing transportation costs and allowing for more stability in the physical and chemical characteristics of the plastic pellets, such as heating value, uniform particle size, and density. This stability allows for a variety of plastic waste materials to be used in the same gasifier. Both processing and type of feedstock plays a role in determining RDF quality. When creating refuse-derived fuel, suppliers should consider the following characteristics in their plastic waste specification

- Well defined calorific value for the plastic waste materials
- Low chlorine content in plastic waste materials

- Quality controlled composition (i.e. plastic waste materials should have few impurities)
- Defined grain size for the plastic waste pellets
- Defined bulk density for the plastic waste
- Availability of sufficient plastic waste quantities with required specifications

The production of ash during plastic energy recovery may add the cost associated with hazardous waste removal and the presence of chlorine may result in the formation of dioxins, which is an environmental concern. Therefore it is highly advisable to separate food residues from plastic waste materials that are sent to energy recovery in order to create drier and lower-chlorine-content fuel. However, refuse-derived fuel approach is most sustainable approach with the pelletisation of plastics for fuel.

KEY STRATEGIES USED IN PLASTICS WASTE FOR ENERGY RECOVERY

Plastics industry has been advocating full valorisation of plastics waste for years and calling for the banning of landfilling of this waste.

Recycling is high in waste treatment hierarchy but not always is it the most sustainable way of waste recovery. For example, multilayer films used in food packaging protect food in an excellent way keeping it safe and extending its shelf life. Till now no reliable industrial process of sustainable recycling of such films has been developed. Another example of a difficult-to-recycle case is plastics material compounded with other materials like glass fibres, metals etc. The share of plastics waste that may be sustainably recycled depends on the nature of plastics products, their application area but also on the existing local waste management system and available infrastructure.

The Austrian consulting company Denkstatt GmbH has tried to estimate limit levels of sustainable (eco-efficient) recycling analysing different plastics items. Denkstatt arrived at a conclusion that with present technology and existing waste management systems actual maximum levels of eco-efficient recycling vary between 29 percent and 45 percent.

Fortunately, such difficult-to-recycle streams of plastics waste are characterized by their high calorific value and may be a perfect component for any fuel from waste. Here we speak of direct incineration of mixed waste (Municipal Solid Waste Incinerators, MSWI) but also of preparations of fuels made of waste (RDF, SRF) that are mainly used in co-incineration processes – e.g. along with coal.

Plastic Waste Incineration

Incineration is the controlled burning of wastes at high temperatures in a facility designed for efficient and complete combustion. By definition, complete combustion involves the conversion of all carbon to carbon dioxide, hydrogen to water, and sulphur to sulphur dioxide. The by-products of incineration are ash, gases, and heat energy. Wastes are burned for one or more of the following reasons:

- Volume reduction
- Destruction of certain chemicals or alteration of chemical
- Destruction of pathogens or
- Energy recovery

Since one of the by-products of incineration is heat energy, the incineration system can be combined with a resource recovery facility where the heat energy can be used to heat water for domestic homes. According to World Bank (2005) report, it is however expensive to set up combustion facilities and the combustion process emits toxic fumes which can only be moderated by expensive air purifiers. The ash from incinerators is loaded with metals and harmful compounds, which should be disposed in secure landfills. Incineration without resource recovery leads to wastage of energy and materials unless it is augmented with a recycling and resource recovery facility.

Fuel from Waste

Refuse-derived Fuel (RDF) or Solid Recovered Fuel (SRF) is a fuel produced by shredding, mixing and dehydrating municipal solid waste and other high calorific waste.

- Generally, all kinds of combustible waste may be used. These include
- Municipal mixed wastes
- Plastic packaging waste materials

- Post-producer plastic waste
- Plastic paper and textiles waste
- Refuse-derived Fuel in most parts of the world can be classified mainly under several categories, which include
- Combustible waste– alternative fuel)
- Plastics and rubber from mechanical waste treatment
- Other non-hazardous wastes from mechanical treatment
- Pre-Refuse-Derived Fuel (RDF) –calorific waste usually coming from Mechanical Biological Treatment (MBT) processes that is stored for future use in RDF formulations

As a consequence of overcapacity of existing MBT plants and increasing restrictions of landfilling processed plastic waste, it has become recently quite popular among waste operators to declare and gain permits for production of alternative fuels. There is a situation where the declared production capacity of alternative fuels is 4 times higher than actual consumption in most countries around the world. Of course, only a part of these capacities (probably less than 50 percent) is capable of producing high quality reproducible RDF that may be sold on the market (both locally and internationally). The other part, usually pre-RDF, is mainly stored at waste operators' premises. Statistically, there is hope worldwide that the alternative fuel market will grow progressively. This is achievable through commissioning of new incinerators and RDF-fuelled power plants.

For instance, in Europe, the largest consumer of RDF is by far the cement industry. Co-firing of RDF in the energy sector is still almost non-existent despite many attempts made to introduce RDF also into this sector. The main obstacles are stricter emission standards required for plastic waste incineration and inadaptability of existing older boilers to co-firing RDF with coal. This situation is slowly changing and new projects of co-incineration of RDF with traditional fuels are currently in the testing phase in Europe and other developed countries such as U.S.A. in Poland, there are two new alternative fuels power plants (Zabrze and Olsztyn) that should start within 2 to 3 years that will use RDF as fuel.

Most cement plants around the world are important consumers of fuels from plastic waste substituting a part of fossil fuels, coal or gas. For instance, cement industry in Poland, one of the most modern in Europe with 10 plants and a total production capacity of 22 Mt/year of cement, has made important investments to adapt their plants for reception, preparation and co-firing of alternative fuels. Those adaptations include storage, drying lines and feeding lines for alternative fuels. Refuse-Derived Fuel constitutes over 80 percent of alternative fuels used in different countries around the world.

Plastics to Fuel/Plastics to Chemicals

For the last few decades, plastics-to-Fuel processes have become some popular in most parts of the world. Production of combustible fuel out of waste has being exempted from excise tax. The conversion of plastic waste materials to fuel is realized through pyrolysis process (thermal cracking of plastics to produce fuel as an output mixture of hydrocarbons that is used as heating oil).

Attempts have been made worldwide in order to market Plastics-to-Chemicals technology with plastics waste as an input and chemicals (solvents, paraffin and waxes) as an output.

There is a visible interest from different state governments across the world who seeks an alternative to management of plastic waste materials through energy recovery processes since such units may be smaller and cheaper to construct.

Plastic Waste Gasification

Gasification is a thermal process that has been in use since the 1800s. At that time, coal and peat served as the gasification feedstock to produce town gas that was piped into cities and homes for lighting and cooking. During World War II, small-scale gasifiers were developed to power vehicles, boats, trains and electric generators. Demand for town gas was eventually replaced with electricity and natural gas.

It was not until the 1980s that management of solid waste materials was used as a feedstock for gasification. According to Engineering-Center for Environmental Research and Technology, (2009) report, the first commercial sized gasification system through the use of solid waste materials was installed in Japan in 1979 processing 100 short tons per day. By 1980, Germany installed a 154 tons per day facility using pyrolysis to process solid waste materials and sewage sludge, also known as municipal sludge.

Energy recovery from waste has been recognized as an essential part of a sustainable integrated waste management system. It has been found that gasification can meet environmental air quality standards when implemented properly. Also, communities with energy recovery systems as part of their waste management plan achieve higher rates of waste recycling (such as plastic wastes). The major advantages of gasification are:

- Reduced land use, 100 times lower than required for landfilling over 10 years
- Modular design allowing for gradual scaling and faster installation than incineration
- Suitability for smaller applications

Production of syngas as a secondary product, which allows for intermediary clean up before it is converted into final energy products.

Gasification Processes: Gasification uses a thermal process to capture chemical energy from feedstock. There are three main classifications of gasification processes, characterized by the reactor temperature.

Pyrolysis- this occurs at lower temperatures between 500°C and 800°C and its distinguishing characteristic is that feedstock is processed in the absence of oxygen.

Conventional gasification-this is thermal processing at temperatures between 550°C and 1600°C, depending on the air mix used in the reactor chamber

Plasma assisted gasification or plasma arc gasification- this process uses a plasma field of electrically charged gas to reach temperatures of 4,000°C to 7,000°C. At these high temperatures, plasma arc systems experience more feedstock flexibility, achieve higher thermal efficiencies, and produce cleaner syngas.

Some conventional gasification systems require that waste is pre-processed to achieve more even feedstock heating and higher conversion efficiencies. Pre-processing of solid waste materials can include removal of un-reactive materials like glass and metal, shredding to create a smaller, uniform fuel material, drying to remove moisture, or pelletization.

Processed solid waste materials are referred to as residual derived fuel (RDF). A diagram of a conventional gasification process is shown in the figure below.

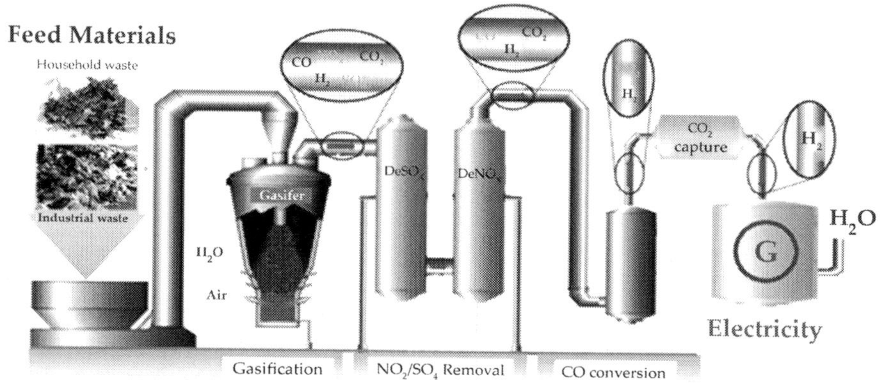

Source:http://cheed.nus.edu.sg/stf/chewch/group2014/biomass_gasification.htm

Figure 9.1: A typical conventional waste gasification system.

Prepared residual derived fuel (RDF) enters the conversion chamber where it is heated to release synthesis gas, which is primarily composed of carbon monoxide and hydrogen. This syngas is cooled and cleaned before being converted into liquid fuels or combusted to produce electricity and heat.

In a typical gasification system, the prepared waste or RDF is fed through a drying chamber and then into a pressurized, oxygen limited conversion chamber and heated from an external heat source. High moisture levels in feedstock lead to increased production of tars at the reactor outlet. This thermal decomposition process produces synthesis gas, also called syngas. The direction of material flows and heat distribution designate different gasifiers. Down draft or up draft gasifiers with stationary or fluidized beds are just some of the available conversion chamber configurations.

The original volume of the feedstock is reduced by 75% to 95% and the remaining solids exit the conversion chamber as char or vitrified slag depending on the temperature of gasification.

Both char and slag are marketable materials which can be repurposed for industrial processes like cement production or as aggregate for other construction materials.

Syngas is composed of a combination of carbon monoxide and hydrogen with the presence of methane, carbon dioxide, nitrogen, and trace contaminants. Syngas is collected from the conversion chamber to be cleaned and cooled before it is converted into secondary forms of energy like electricity or liquid fuels. According to U.S. Department of Energy, (2012), Syngas has a similar composition to natural gas and can be combusted in a combustion engine, Brayton cycle, or Rankine cycle to produce electricity. The other common utilization of syngas is conversion into liquid fuels via a Fischer-Tropsch reaction to produce an alternative to conventional transportation fuels. Additionally, new studies are identifying ways to use syngas to generate electricity in fuel cells.

Whilc product gases from gasification can include air pollutants such as particulate matter, aerosols or tars, nitrous oxides, sulphur oxides, dioxins and furans, and hydrocarbon gases there are multiple points of intervention available for controlling emissions because gasification produces a secondary energy product. Gasification systems can employ pollution control systems at the reactor outlet as well as the exhaust gas outlet.

A 2009 study produced by the University of California, Riverside determined that the pyrolysis and gasification facilities, currently operating at the time, met emission limits mandated in California, the United States, and the European Union with only a few exceptions. In every gasification process that was evaluated, levels of toxic air contaminants of dioxins, furan and mercury met the most stringent standards worldwide.

Plastic Pyrolysis

Pyrolysis takes place lower temperatures in the absence of oxygen and produces intermediate products of syngas, pyrolysis oil, and char. Pyrolysis has been used extensively in manufacturing to produce products like coke for steel processing and industrial chemicals. A drawback of this process in the high level of pre-processing required when treating municipal solid wastes. Uniform densities, feedstock sizes, and moisture levels greatly influence the conversion efficiency of these systems.

There are two main types of pyrolysis processes, "slow pyrolysis" and "flash pyrolysis". Slow pyrolysis takes place in a stationary reactor and used to maximize the production of gas and solid by-products. Flash pyrolysis occurs over several minutes and is used primarily for producing pyrolysis oils.

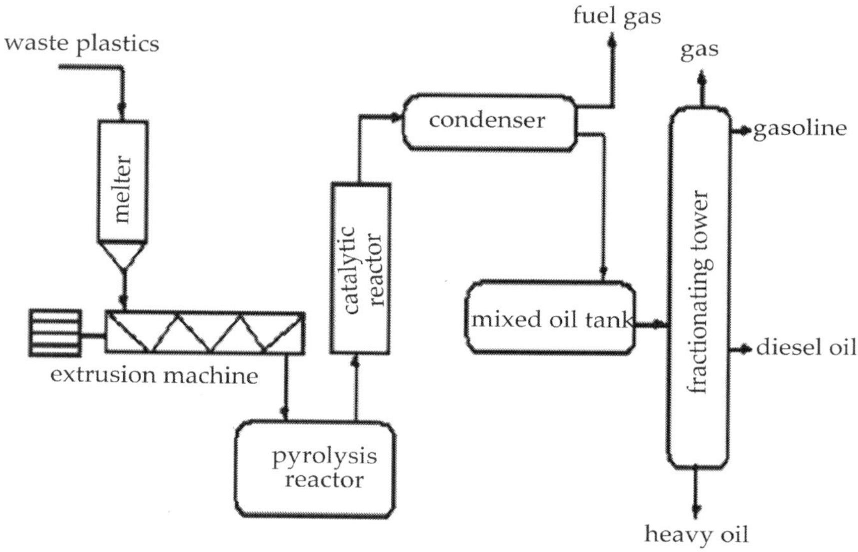

Source:https://www.researchgate.net/figure/237008261_fig3_Figure-3-Pyrolysis-catalytic-upgrades-technique-of-plastic-wastes

Figure 9.2: Pyrolysis-catalytic technique of plastic wastes.

Syngas from pyrolysis contains methane, as well as carbon monoxide, hydrogen, carbon dioxide and water. Pyrolysis can produce high amounts of tar which damages generation equipment and clogs syngas filters designed to remove other pollutants.

The char and ash produced from pyrolysis can have varying levels of toxicity. Disposal of char from pyrolysis ranges from combustion as a secondary heat source, repurposing as a cement additive, to specialized disposal in a mono-fill.

Plasma Arc Gasification

Plasma arc gasification is the newest development in waste gasification technology. Also referred to as plasma assisted gasification, plasma produced from electricity provides the heat source for the gasification process. Plasma is a stream of high voltage, high current electricity produced from a plasma torch, and has characteristics similar to lightening.

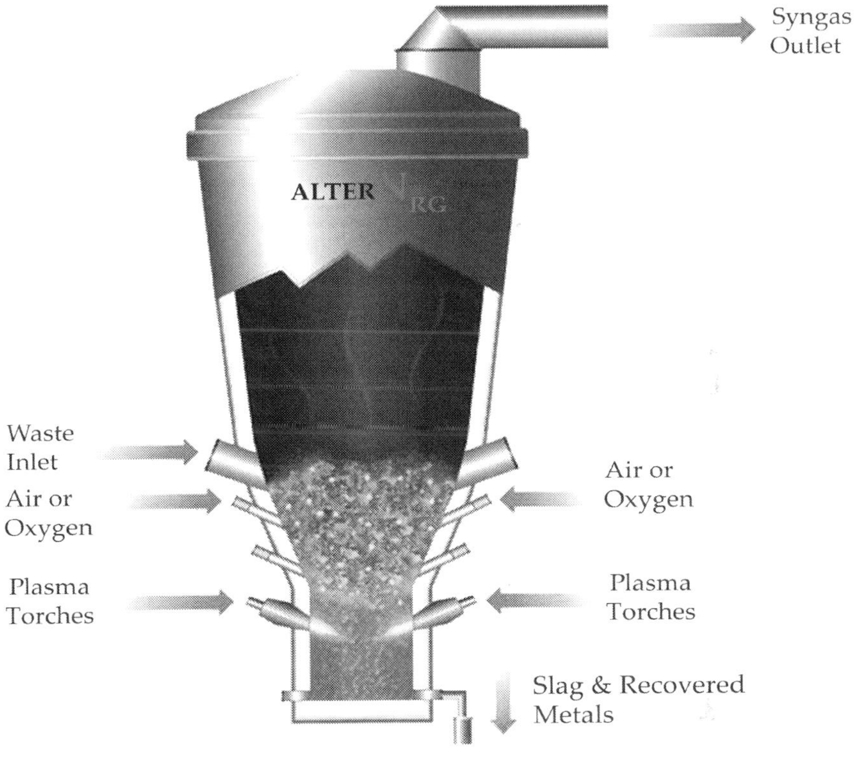

Source:http://www.gasification-syngas.org/technology/plasma-gasification/

Figure 9.3: A typical Plasma Arc Gasification setup.

The plasma field reaches temperatures up to 14,000°C. Even a few feet from the torch temperatures can be as high as 2,700°C to 4,400°C. These extremely high temperatures break down municipal solid waste more completely giving plasma gasification systems increased feedstock tolerance and higher syngas yields than traditional gasification while producing less char and tar. In some systems, syngas is passed through a secondary plasma field for additional cleaning. The solid by-product of this system is vitrified slag or glassy slag, which exceeds EPA leach test standards.

The oldest operating plasma arc facility is located in Japan. It was commissioned in 2002 by Hitachi Metals and uses a Westinghouse Plasma Corporation plasma gasification process. According to Engineering-Centre for Environmental Research and Technology, (2009) report, the largest plasma arc facility has a processing capacity of 274 tons of municipal solid waste per day. Plasma arc gasification is growing in popularity because it has more flexible feedstock acceptability, can produce cleaner syngas, and achieves higher electrical efficiencies compared to conventional gasification.

10 ADEQUACY AND ENFORCEABILITY OF LEGISLATIONS AND POLICIES WITH REGARD TO PLASTIC WASTE MANAGEMENT

INTRODUCTION

Plastics wastes have been a prevalent issue for a number of decades across the world. The increase in plastics pollution can be attributed to the high population growth and industrialization in a country. For instance, plastic packaging bags are issued by supermarkets and at the market place due to their high convenience, strength and inexpensive nature as a reliable way to deliver goods from the store or market to the home of the customer. The high population growth contributes to the problem of pollution in that there are many households and each household uses plastic products in the form of plastic bags for the carrying and packaging of goods. After the plastics have achieved their purpose they are thrown away as waste some of which remain uncollected or discarded in a disorderly fashion. This leads to an increase in plastic waste pollution which can carry serious health risks. Since the 1980 there has been an increase in the rate of which sewage drains have been clogged all this due to the prevalence of plastic waste materials that find their way to the sewage systems. This causes a lot of problems, the nasty smell and unsightly fluid but the worst is that it causes diseases such as cholera. The clogging of sewerage drains also increases the chance of creating a haven for mosquito breeding ground which in turn leads to the spread of diseases such as malaria.

Most of the plastic materials used in packaging various products end up in landfills and this contribute to land pollution as plastic waste materials take a long time to decompose due to their non-biodegradable nature and this causes the inhibition of biodegradable material in or around the plastic materials hence affecting the fertility of the surrounding area.

Plastic waste pollution can also contribute to the deaths of many animals and birds alike. This is due to the fact that when an animal ingests the plastic waste materials, it gets choked leading to suffocation or indigestion which leads to the animal or birds death. The plastics menace also causes most cities that were once referred to as the "city in the sun" in those heydays to lose all of their glorious beauty which in turn costs tourists who are very vital for a country's economy.

Various studies have been done on the subject of waste management for instance studies on the legislative and institutional frameworks governing solid waste management.

For instance, a country like Kenya has a legal framework on waste management comprises of the Environmental Management and Co-ordination Act of 2015 (EMCA). This legal framework acts as the principle guideline setting up a foundation for other environmental laws. It regulates environmental policy in Kenya and establishes the National Environmental Management Authority which is the institutional body concerned with environmental matters in Kenya. The EMCA also has regulations that address the environmental problem at hand for instance the EMCA Waste Management Regulations of 2006 which deal with the management of solid waste and provides for regulation on the collection and disposal of solid waste products in the country. It also provides for the proper management and maintenance of areas where waste material is discarded. This is to prevent people who are not authorised from accessing areas that are harmful and hazardous to their health. It also comprises of health and sanitation Acts involved with waste management control in Kenya such as the Occupational Health and Safety Act of 2007, which deals with proper waste management practices with a view to curbing the spread of diseases caused as a result of poor waste management practices. The legal framework also incorporates by-laws formulated to help in the curbing of poor waste management practices for example Nairobi County has two laws relating to the management of waste as well as the production use, the Nairobi City County Solid Waste Management Bill 2015 and disposal of plastic bags the Nairobi City County Plastic Carry Bag Bill 2015.

But from a broader perspective of health, the government of Kenya had to act and it rightly banned the use of plastic products with some exemptions as the first step of a broader vision to eliminate the use of plastic in day to day life. But then, was this the right way to go about it? Was the general public willing and ready to come out of their comfort zones? Could the government of Kenya have reduced the usage of plastic through better methods like penalizing the usage, incentivizing the alternatives like paper bags and cloth bags? These are some of the key questions that every government with plans to ban plastic materials should ask first before initiating a banning program. This is because most countries around the world are heterogeneous states with varied consumers and extremely diverse behaviour patterns.

Validation of the Study

Research studies have revealed a country may have some loopholes in the law that allow for the continued plastics usage and plastic pollution. These loopholes may lead to a continuing pile up of plastic waste in landfills. Furthermore the increase in the companies that manufacture and produce plastic products may also allow the increase of plastic waste in the environment.

To effectively curb the menace of plastic waste pollution in environment through enactment and enforcement of laws, there is a need to conduct a research on the impacts of relevant laws governing plastic waste pollution on the environment and how they should be implemented. Plastic waste pollution is a serious issue that affects all flora and fauna in the ecosystem.Given its non-biodegradable nature it could cause a lot of problems among them enhancing the spread of diseases, getting into the food chain, blocking sewage channels and spoiling the general aesthetics of the country. Therefore, this problem can be regulated and managed through enactment of effective laws. Additionally, there is a need to study various policies that can be implemented to ensure that plastic waste pollution is controlled and ultimately eradicated. According to Wangari Mathai (Nobel Peace Prize, 2004) quote;

"*You cannot protect the environment unless you empower people, you inform them, and you help them understand that these resources are their own, that they must protect them.*"

There is also the need to research on ways that can be used to change the attitude of people towards maintaining a clean and pollutant free environment through public participation. Another recommendation entails studying how environmental impact assessment reports with regard to plastic waste pollution can be implemented and the policy of reduce, reuse and recycle can be adapted to help in the fight against plastic waste menace in the environment. Finally a study is required to brainstorm on the notion of regulating, managing, reduction and ultimately non-reliance of plastic products.

ENVIRONMENTAL LAWS ON PLASTIC WASTE POLLUTION

Environmental law has its origins in environmentalism which is a value system that seeks to redefine the relationship of mankind and nature and to induce man to be stewards of nature rather than exploiters of nature with a view to respecting the functioning of natural systems and by limiting activities that disturb these systems. According to ecological land ethics by Aldo Leopold "a thing is right when it tends to preserve the integrity, stability and beauty of the biotic community. It is wrong if it tends otherwise." This land ethics is the leading land use management alternative to the progressive conservation movements strategy of multiple use of resources and has provided the basis for power resource preservation and protection laws. The land ethic has for a long time been used by many legal scholars to formulate different environmental regulation laws mainly because it follows the equilibrium paradigm (balance of nature) because it was seen to be a neutral organizing principle that could be applied to the use and management of all natural resources. The equilibrium theory which is based on a balance of nature sees the need to let nature regulate itself and is of the view that man should let nature be. In other words, equilibrium theory is usually against the interference of man in nature.

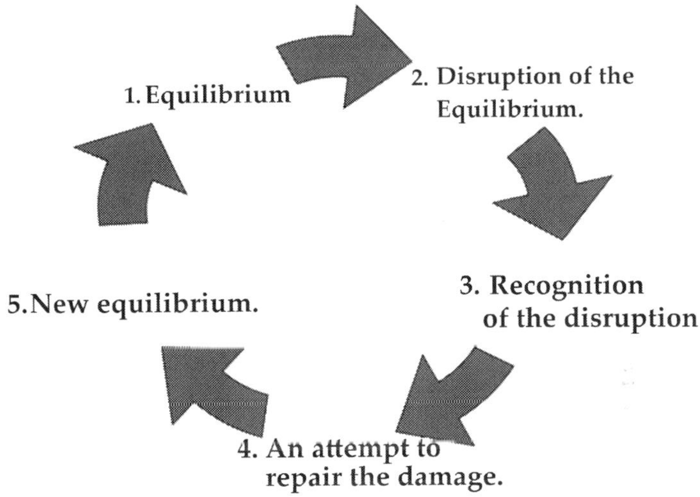

Source:https://www.slideshare.net/Pinkgirlchloe/todorovs-theory-28049389

Figure 10.1: Todorov's equilibrium theory.

The equilibrium paradigm has a large basis for its justification and thus can accommodate new views of environmentalism. However it is based on ethics rather than science thus it is possible for the equilibrium paradigm to have flawed interpretations that arise due to passions rather than logical and reasoned out thought. The equilibrium paradigm does not focus its attention on the primacy of science which overtime has become one of the bases of formation of environmental laws. The fact that it relies primarily on ethics makes it a difficult choice to use it as a theory to justify the need for an effective and adequate regulatory framework for plastic waste pollution control. This is because in order to make laws that regulate polythene use, one must consider the science behind it in order to make informed decisions on the best applicable laws. Secondly the equilibrium paradigm is of the view that nature should return to its former state free from human interference. This means that human laws should in no way affect biodiversity as this would amount to interference of nature and this means that man has been left unregulated to interact with the ecosystem. A little interference in the ecosystem through regulation of man's relation with nature is needed to ensure the protection of the environment. These ties in with plastic waste pollution control in that man should not be allowed to use the ecosystem as a dumping site; the law should regulate the actions of man to prevent him from causing environmental pollution of any kind given that plastics can affect the ecosystem in multiple ways.

Environmental law derives its political power and legitimacy from science. Ecological and toxicology have identified a large range of harm to be caused by human factors. The same science that environmental law derives it political power and legitimacy, is the science that can remedy the harm caused by human factors. Environmental law divides itself into two broad fields; that is pollution risk remediation which is the correction of human caused problems affecting the environment and biodiversity protection.

The discussion on this chapter is based on the theories of environmentalism, with a keen interest on the non-equilibrium paradigm. The Non-equilibrium paradigm supports the position of the prevention and mitigation of human disturbances that affect both land and water ecosystems with a bid to protect biodiversity from destruction. The non-equilibrium paradigm dominates ecology and toxicology which are sciences that provide for justifications a wide range of prohibitions on human activities that alter the land and water ecosystem, they form the basis on which pollution control laws are made.

The Non-equilibrium paradigm rejects the notion of a balance in nature, which is supported by the alternate paradigm of equilibrium. The equilibrium paradigm on the other hand postulates that nature should be left alone and man should not interfere with nature. Scholar maintain that it will be difficult for the equilibrium paradigm to work as there is constant human interaction with nature and this makes it impossible for the ecosystem to revert back to the ideal state of nature.

The Non-equilibrium paradigm shares the view that there is a need for biodiversity protection and that unregulated humans pose a threat to biodiversity.Ecologically, there is more room for scientists to come up with possible solutions and remedies for ecosystems that have already been damaged by human disturbances through the paradigm's constant need for research. Through this approach legislators can come up with laws and regulations that limit human disturbance and keep it to a minimum.

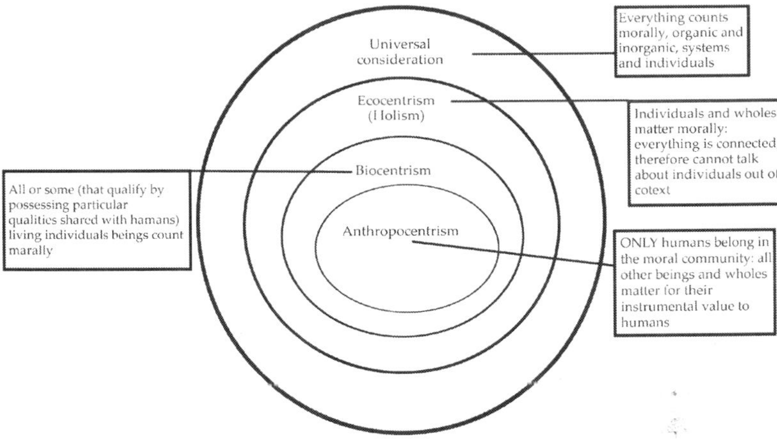

Source:https://www.researchgate.net/publication/268506880_Anthropocentrism

Figure 10.2: Illustration Anthropocentrism theory.

In relation to the Non-equilibrium paradigm, the Non-anthropocentric theory of environment is helpful in ensuring the protection of biodiversity and its ecosystems. The Non-equilibrium paradigm pushes for the rights of nature as opposed to the balance of nature this augurs well with Non-anthropocentric views as man is removed from the centre of the ecosystem and the focus lies entirely on the environment as the main interest that needs protecting. The laws are geared towards the protection of the environment as it is not being used by man alone but by various flora and fauna. The Non-equilibrium paradigm does not undermine the need for biodiversity protection. This means that this paradigm understands that there is a need to protect the ecosystem as a whole and this notion greatly supports the non-anthropocentric theory whereby all focus of the environment is drawn away from man and focus is placed on the ecosystem as a whole.

It is of utmost importance that when creating laws involving environmental regulation there is a need to focus more on the environment and less on man. This does not mean that we ignore man's actions that may impact the environment, it is imperative that man's actions be checked because if man is left unregulated, he can cause a lot of damage to the ecosystem and the magnitude of human integration is always too great.

The Non-anthropocentric theory also encourages the use of adaptive management strategies which are management strategies that change with in response to new scientific information. This works well with the non-equilibrium paradigm of using science as a basis of legitimizing environmental law and also shows that there will always be a new way of doing things and that is why there is a need to adapt a mechanism that changes with time.

The Non-equilibrium paradigm and Non-anthropocentric theories work well with plastic waste pollution regulation because the Non-equilibrium's reliance on sciences such as ecology and toxicology as a basis of justification offer it the necessary information required to be able to formulate laws that will better preserve the environment. An informed research on the causes of plastic waste pollution on the ecosystem leads to consideration of the impact on environment that the pollution will have rather than focussing solely on the human impact. This brings about the aspect of Non-anthropocentrism as the focus will be on the entire ecosystem and not just on man. In making regulations it is imperative to know the effect those laws may have for instance a plastic waste pollution regulation may propose that all plastic waste be dealt with by burying them in the ground or by burning them, this would be a good regulation as it will have dealt with the plastic waste pollution problem. However, it does not bear in mind what those actions might have on the ecosystem and the overall environment. This is where the science in ecology comes in, by opting to burying the plastic waste in the ground; it affects the fertility of the ground as polythene is a non-biodegradable material and will also affect other biodegradable material necessary for soil fertility. By opting to burn the polythene will result in the formation of toxic fumes that will pollute the air and might contribute to health complications to the people living in the surrounding areas.

Different studies show the use of science as a basis for reliance in coming up with environmental regulation. They also show non-anthropocentric views through considering the ecosystem such as the soil and the air in making of environmental regulations.

According to scientists, there has been a growing concern with regard to the state of the society, they argue that man has seen himself as the lord of the Earth and is plundering the earth at will and this violent nature of plundering the Earth is evident in the sickness that present themselves in the air, on land, in the water and in all forms of life. Scientists have also pointed out that man is at the apex of destroying the environment of which he is part of, as he is made up of the dust of the earth and is composed of the elements of the earth through the air he breathes and receives life and the waters that refreshes him.

Pope John Paul II also touched on the issue of the dilapidated state of the environment in his Encyclical whereby he warned that human beings seem to see no other meaning in the environment other than what serves their immediate use and for consumption. In the same Encyclical Pope John Paul II called for a global ecological conversion after noting that little had been done to safeguard the moral conditions for an authentic human ecology.

The Encyclical letters of the Pope counters the views of many scientists that religion is a cause rather than a solution to the environmental problem. This is because through the edicts of the Holy fathers, they show a great concern to the state of the environment and they attribute this problem to men who in their actions destroy their environment without thinking of the consequences that might have on them and on future generations.

According to World Vision (2011) report which aimed to show a relationship between man and the environment focussed on the view as man as a steward for the environment, this moves away from the non-anthropocentric theory as it places man in the centre of the environment but instead of being the antagonist, man is given the role of a steward, a protector and carer for the environment.

They borrow from the Bible whereby Man is given a responsibility by God to work and take care of the Garden of Eden which can be equated to the environment

This is a fresh and biblical way by which man's action on the environment can be examined. Using such an approach in the protection, conservation and preservation of the environment could prove to be beneficial given that man uses the environment's resources for his own survival and he does not use it alone and thus must be ready to prevent any incidences that might cause harm to the environment and this includes preventing plastic waste pollution given that it affects multiple aspects of the environment.

The Non-anthropocentric theory ties in with the research as it removes the focus away from man as being the apex of the environment and thus disallows him from plundering the earth as he wishes. This is attributed to the fact that man does not use the earth's resources alone and his actions have consequences. Through the lack of adequate checks and balances man has been able to choke up the earth with plastic products.

The notion of using man as a steward for the environment is also helpful in tackling plastic waste pollution as it gives man a responsibility over the environment and as such he is required to care for it, to protect, conserve and to preserve it as mandated by God in the Bible. However, it differs with the non-anthropocentric theory as man is still the main focus of the environment but it comes around to assert more responsibility on him for being the apex of the environment. Using the two approaches could therefore offer the possibility whereby man has responsibility and is a steward of the environment but he is not at the main focus of the environment. He would be in the position that he shares the environment with other flora and fauna with the ecosystem being the main focus of the environment.

CURRENT STATE OF THE ENVIRONMENT IN THE WORLD WITH REGARD TO PLASTIC WASTE POLLUTION

The state of the environment presently is that there is a large amount of plastic products littered all over the world and especially in major towns and cities. This has greatly been caused by the lack of effective environmental policies with regard to plastic waste pollution and weak regulatory frameworks to aid in the eradication of plastic waste pollution.

One problem facing plastic waste pollution control and eradication is that there is a lack of demarcation as to what constitutes plastic waste and what differentiates it from other types of solid wastes. Most the Environmental Acts do not mention plastic waste as part of solid waste, they generalizes the types of waste and puts them all in one category of solid wastes. This lack of demarcation between solid waste and plastic waste puts a strain in the efforts of having a plastics free environment given that there is no long term solution to the problem of plastic waste pollution.

Waste management in general is a huge problem in most parts of the world and it is attributed to the increasing rate of urbanization and rural-urban migration. The rapid urbanization has led to an increase in the production of solid wastes as there are more people and the higher the number of people the higher the number of waste products they produce. The problem comes in when the rate of solid waste collection is lower than the production of solid waste.

For instance, according to CCN (2007) report, in 1992, 800 to 1000 tonnes of solid waste were produced in Nairobi (which the capital city of Kenya, in Africa) alone and less than 10 percent of the waste products were collected, in 2002 the solid waste products increased to 1530 tonnes 40 percent of which remained uncollected or disposed of carelessly or burnt. Currently the rate of solid waste generation in the same city is 4,016 tonnes on a daily basis and over 2,390 tonnes remain uncollected and a huge number of the total solid wastes include plastic products.

The state at which plastic waste pollution is rising is worryingly high, as of 2007 over 2 million plastic bags were being handed out as carrying bags and after the use of such bags most of them were disposed of through improper channels and some remained uncollected. The careless disposal of plastic wastes has contributed to a lot of problems in the world for instance it has caused blockages in waterways and drenches contributing to flooding which in turn provides for breeding grounds for mosquitos which spread malaria among other diseases as well as spreading the risk of contracting cholera. The plastic waste also contributes to choking of wildlife and soil pollution.

Source:https://guardian.ng/news/nigerians-decry-the-scourge-of-polymer-products-wastes/

Figure 10.3: Illustrating how plastic waste materials have evaded drainage channels and sewer systems.

The Lesson Learn from Kenya and Rwanda Case Studies

Kenya

Different countries have different environmental laws that regulate environmental issues. Kenya for instance has the Environmental Management and Co-ordination Act of 2015. It was enacted against a backdrop of sectorial laws dealing with various components of the environment, the deteriorating state of Kenya's environment, as well as increasing social and economic inequalities, the combined effect of which negatively impacted on the environment. The supreme objective underlying the enactment of the Act was to bring harmony in the management of the country's environment. This is the main legislation that addresses environmental issues and sets up regulatory bodies that handle environmental issues such as policies, research advisory opinions to the government as well as advising the National Assembly on the laws to be promulgated. Such bodies include the National Environmental Management Authority (NEMA) which is a semi-autonomous government body created under section 9 of the Environmental Management and Co-ordination Act and is mandated to exercise general supervision and coordination over all matters relating to the environment and to be the principal instrument of the Government of Kenya in the implementation of all policies relating to the environment.

The Act classifies hazardous waste under various classifications and their effects to the environment as well as to persons. It defines hazardous waste as any waste which has been determined by the Authority to be hazardous waste or to belong to any other category of waste provided for in section 91. This definition is wide as it makes plastic waste products open to a broader spectrum as plastic waste affects the environment in more than one way. With the lack of a proper definition of plastic waste pollution it makes it harder for efforts addressing the rising problem to be effectively implemented. The act has broadened the spectrum in which some of the solid wastes are to be looked at and in turn has caused confusion in the collection of solid waste leading to degradable waste being collected together with non-biodegradable waste thereby hampering the efforts of recycling or change of use of those substances. The law should be clear and non-ambiguous to allow for better preservation, conservation and management of the environment.

Given that the Kenya constitution under article 69 (2) mandate that the protection of the environment be a duty of all the citizens, county governments such as Nairobi have formulated legislations to deal with plastic waste pollution. For instance, the Nairobi City County Plastic Carry Bags Bill of 2015 is a mode of ensuring that there is an end to plastic bag pollution in the county. The objective and purpose of the bill is to provide for the control of pollution and the conservation of the environment within the precincts of Nairobi City County. The Bill that is yet to come into force seeks to control the manufacture of plastic carry bags and other plastic products from recycled plastics and pigmented carry bags, control the use of those bags and prohibit the disposal of non-biodegradable plastic waste in public places, drains and sewers for restoring and maintaining clean environment. There are a number of parameters stipulated in the bill. For instance;

The Bill prohibits the manufacture of plastic carry bags of less than 30 microns and 8” x 12” in size under section 5 contravention of the said section would amount to an offence.

The Bill also disallows the use and sale of plastic non-biodegradable carry bags by retailers and further disallows the handing out of the carry bags free of charge.

The Bill provides for the prohibition of disposal of plastic bags in public places or in places where there is a public view. It also disallows the disposal of bio-degradable material in non-biodegradable bags.

The Bill also calls for a permit for any person seeking to remove non-biodegradable plastic waste especially if such removal would cause harm to the public or cause any drainage.

The Bill also offers enforcement mechanisms such as a department that will undertake the task of filing reports, training personnel on collection methods; carry out research on proper methods of handling with plastic waste and create public awareness among other duties.

However the Bill is not clear as to which the relevant authority is and the constituent members.

Under section 31 it provides for general penalties for any party found in contravention with the

Bill. It provides that any person found in contravention of the Bill for an offence that no penalty has been laid out will be liable for an imprisonment of up to one year or a fine worth five hundred thousand shillings.

Recently, the government of Kenya took some steps in trying to correct the issue of plastic waste pollution, for instance in 28th August 2017 the government introduced a policy banning the use of plastic bags with less than 30 microns and this has gone a long way in reducing the amount of plastic waste currently in dumping sites and all over the country had it not been greatly opposed by the members of the public. The Environmental Management and Co-ordination Act of 2015 is a useful tool in combating plastic waste pollution given it is the national law that covers the protection of the environment.

The Constitution of Kenya 2010 under article 69(1) gives responsibility to the state to ensure that the environment is well managed and conserved and the natural resources are equitably shared.

This responsibility also applies in the formulation of policies and legislation that would address the issue of plastic waste pollution in the country as provided for under article 72 of the Kenyan Constitution.

The Constitution gives the state a mandate to ensure that the environment is protected against all forms of pollution including plastic waste pollution. So far, not much progress has been made in protecting the environment from plastic waste pollution in a national level but through devolution some counties for example Nairobi have come up with legislation that will help in the efforts of eradicating plastic waste pollution in the country. This is a step in covering the larger part of the country in eradicating plastic waste pollution.

The Environmental Management and Co-ordination Act has established the National Environmental Management Authority (NEMA) which is the regulatory body for the management and conservation of the environment. They carry out surveys, which assist in the proper management and conservation of the environment, advise the Government on legislative and other measures for the management of the environment or the implementation of relevant international conventions, treaties and agreements as well as advise the Government on regional and international conventions, treaties and agreements to which Kenya should be a party and follow up the implementation of such agreements among many other functions. This is the body that tasked with ensuring that the environment is well managed and conserved to ensure that there is no environmental degradation. However there has been a laxity in combating plastic pollution given that the rates are still rising. There is lack of an implementing arm in the body and this might explain the reason as to why there are minimal to no effort in addressing plastic waste pollution in the country.

Rwanda

If Rwanda was in the spotlight of the international media for over 20 years due to the 1994 genocide, today it is mostly owing to its economic resurgence and its development projects, such as Vision 2020. One of its achievements is the prohibition, since 2008, of the production, importation, use and sale of polythene bags on a national scale. The fact that Rwanda has managed to abolish all plastic bags is in itself a remarkable achievement. Based on what some of the studies show, if this ban on plastic bags has been possible in Rwanda, it is because “it comes from the top leadership and when a law is decided, it’s serious”. Many other countries in Africa have tried to implement the same policy, such as Senegal or Tanzania, but have not yet managed. Similarly, in Europe, it seems there is a long way off from a total prohibition on plastic products. A few weeks ago one could read in the Flemish newspaper, De Standard (2015), that all member states of the European Union should work to reduce the average use of plastic bags consumption or that by 2018 the consumer would be required to pay for (except in the case of meat or fish) every plastic packaging bag used. This only concerns plastic bags thinner than 0.05mm thick, also called single use plastic bags and not all plastic bags.

Numerous journalists explain that in a democratic country the banning of plastic bags is a very difficult reform that would require a considerable amount of time to enter into force, while in Rwanda, the authoritarian regime in place facilitated the implementation of the law in an efficient and short period of time. However, even though Rwanda is the first country to push the law so far as to prohibit the manufacturing, use, sale and importation of plastic packaging and carrier bags, a few other cities or countries have banned plastic bags too. As an idea for further research, it would be interesting to do a comparative study on two very different places, in order to analyze how aspects of society-historical, cultural, political, etc. have created an enabling environment for such a ban to occur. The ban on plastic products in Rwanda has been successful due to the connection of several elements such as culture, power and sustainability.

Culture: The role played by women in the aftermath of the genocide was a critical element that contributed to the situation enabling a plastic bag ban to occur in Rwanda. Women representing the majority, after the genocide, played a major part in all aspects of the country's healing and reconstruction phase. Without their hard work, Rwanda would not have been able to develop its numerous projects, such as stipulated in Vision 2020. This process, which changed and transformed the role of women in society, has created two distinct groups of women: a group of women dedicated to traditional roles, and a growing group of women having non-traditional roles.

Behaviors and habits have also been affected by the plastics ban. If Rwandans were once used to go to the market knowing they would pack their purchase in plastic bags, today men and women have to plan when shopping and take cotton bags or baskets to carry their goods from the market. Moreover, the disappearance of wandering plastic materials from the streets impacted the consumer, who consciously adopts a more responsible waste management attitude.

Through the monthly community works, *Umuganda*, Rwandans come together to clean their neighborhoods of plastic waste materials. These gatherings are not only important for the cleanliness of the country but also to bring all Rwandans together for the benefit of their country.

Cultural changes have also been observed in the political will to create a national unity and inclusive ethnicity. However, doubts still exist regarding this national unity, which in the eyes of many is more perceived as a 'tutsization' or pressing of cultural assimilation than an inclusive ethnicity.

Rwandan history has proven how influential ICT, such as radio and TV broadcasting, can be on people's values, ideas and behaviors. During the genocide, this media of communication urged Hutus to kill Tutsis. Today, ICT are increasingly used for broadcasting programs to raise awareness regarding the environment.

Lastly, there appears to prevalent reluctance among the population in general and chiefly among those suffering from the plastic ban to openly talk about 'power'.

Power: The political regime in Rwanda, an authoritarian one lead by a Tutsi president, is a regime that leaves little to no room for critique. A handful of people exert decision- making power while the majority simply obeys. Even in the political arena, where Rwanda ranks as the country having the highest percentage of women in the world, women have little say in political decisions. In addition, political opponents are being hunted down, while Hutus are near to absent of the political scene.

Furthermore, different forms of power are being utilized to maintain compliance, namely law enforcement, punishment and discipline. The REMA, in charge of the law enforcement, controls that nobody enters the country in possession of plastic products. The ones caught, face high risks of severe punishment. Being aware of those consequences, the Rwandan population disciplines itself so as not to be confronted to justice.

Sustainability: Growing numbers of political opponents and human rights activists have fled the country these last years and have not come back. This is mainly owing to the fact that there is no room for critique and the fear of being punished. It is obvious that fear and discontentment does not ensure political sustainability.

Today ethnical divisions are still central and unlikely to disappear in a nearby future. If the government were able to work on equality regarding roles and responsibilities without disregarding cultural difference, this would allow for cultural sustainability. However, imposing a dominant culture would only weaken this cultural sustainability, as history has proven how often this causes conflicts.

Lastly, to use the sayings of Tutsi exile former Chief of Staff to the current Rwandan president, Theogene Rudasingwa; "Rwandan exiles are still very concerned and live under constant threat of the opposition. For there to be sustainable prosperity for people, you have got to enrich people's right."

Evidently, this shows how through the plastics ban, multiple aspects of a society can be studied. This certainly was the case of Rwanda, where it was possible to go as far as to understand the impacts on and of the political regime, gender equality, national unity, and cultural assimilation amongst many other elements of and on the plastics ban.

LAWS AND POLICIES IMPLEMENTATION TO REDUCE PLASTICS CONSUMPTION

It is important to have a proper policy framework in place in order to effectively reduce consumption of plastic product. Studies have shown that good communication to the general public and implementing a tax before a ban on plastic products provides the best result in decreasing consumption. In other words, there should be a slow transition in order to be successful in reducing plastics consumption in a country. There should be clear communication to the general public about new laws being implemented before the rules are put into place to allow gradual change in people's behaviour. It is worth noting that issuing a ban without enough clear information creates friction with the public and non-compliance with the law. Therefore, the following are steps on how the responsible authorities can reduce plastic consumption effectively:

Communicate and Implement a Tax per Single-use Plastic Ban Sold

Stakeholder consultation and communication: The first step is to have a consultation between all the stakeholders involved. Some of the key stakeholders in this situation it would include;

- Retailers
- Supermarkets
- Restaurants
- Clothing stores
- Local law enforcement
- Consumers
- Environmental organizations
- The local government

Having a consultation first will allow the stakeholders to come together and address the plastic consumption issue. Deciding a way to decrease plastics usage together will minimize noncompliance once the regulation is in place. After the consultation, the next step is to communicate the new regulation to the public through a variety of ways. Local communication such as radio and television commercials, as well as signs hung in businesses should be utilized. It must be mandatory for every business to put up these signs in the front doors and at cash registers. The sign must be visible, undamaged, and not covered in any way at all times. Tourists should be informed of the new regulations in points of entry like airport and marine ports. The messages should advertise the regulation and why reduction in plastic usage (especially single-use plastic bags) is important for a country's economy.

Implementation: After about six months of communication the government should evaluate the previous process for strengths and weaknesses and then determine if a tax can be implemented. If so, the government should first develop a system to disperse the money collected then determine the tax amount. Based on different previous studies analyses it can be recommended that a $0.15 tax per single-use plastic packaging bags is implemented at every supermarket, grocery store, restaurant, take-away food stand, department, and clothing stores for at least one year.

To change attitudes toward plastic products and to reduce consumption, implementing a tax will make consumers think twice before going to the supermarket without bringing a bag. There are other countries and states such as Ireland and Washington DC, USA that have had success in decreasing consumption using a tax. For instance, Ireland implemented a 15 euro- cent tax in 2002, which is about $0.17 in US currency, and has reduced consumption by 90 percent. Washington DC has a $0.05 tax but this charge is too low and has not changed consumer behaviour. Even though the tax in Washington D.C. has not drastically changed consumer behaviour, the money collected from the tax usually go towards cleaning up the environment. Governments from other countries can do the same with the tax implemented. Allocation of tax money can be done in order to provide education and information to the general public, support workshops and other environmental efforts in reducing consumption of various plastic products. In addition, the government can use the money to possibly create jobs to make reusable bags and other bio-degradable packaging and carrier materials. The reusable bags could be sold in bulk to retailers to be distributed at places like supermarkets and can also be sold at markets for cruise ships to tourists that want souvenirs from a country.

Communicate and Implement a Ban on Distribution and Use of Plastic Packaging Products

Stakeholder Consultation and Communication: Just as mention above on plastic products taxation, the first step is to have a consultation of a ban between all the stakeholders involved. The consultation would involve the same stakeholders; retailers, supermarkets, restaurants, clothing stores, local law enforcement, consumers, environmental organizations, and the local government. Having a consultation first will allow the stakeholders to come together and address the plastics products consumption issue. Deciding the best way to ban plastic products together will minimize noncompliance once the regulation is in place. The next step is to communicate the new ban that will be implemented. Communication should target locals and tourists. Information for locals should be disseminated through radio and television commercials and signs hung in businesses. It must be mandatory for every business to put up these signs in the front doors and at cash registers. The sign must be visible, undamaged, and not covered in any way at all times. To inform tourists of the ban, short videos need to be shown to incoming flight passengers of the new legislation and large signs should be put up at the airport and areas where cruise ship passengers enter a country. In addition, hotels should provide information by using signs in every hotel room to reinforce the ban. Again, the message should advertise the ban and why reduction in plastics usage is important. Once the ban is put into place the taxation law ceases to exist.

Regulation and Enforcement: The ban should prohibit the use and sale of plastic products and control importation to the country. This should include all packaging plastic containers and plastic packaging bags distributed by supermarkets, grocery stores, restaurant, take- away food stands, department, and clothing stores.

There needs to be sufficient regulation and enforcement in order for this ban to be effective. A country like Haiti was not successful in decreasing the use of plastic products, especially the single-use plastic bags because it had little to no regulation and enforcement of its ban. Learning from Haiti's failures, any country should have efficient regulations and enforcement to achieve a significant decrease in plastic consumption.

Challenges in Implementing Laws and Policies on the use of Plastic Products

There are several aspects that determine a country's success or failure in reducing plastic products usages. The following three factors are significant in reducing plastics consumption

- Communicating information to the public
- Enforcing policies and regulations
- Providing suitable alternatives to plastic products

As long as the government has these three aspects in mind when creating a tax or ban, a country can be successful in its ultimate goal of reducing plastics usage.

Communicating information to the local public may pose to be a difficult task because of language barrier within a country with several tribes. For instance, a country like Kenya has more than 42 tribes. This array of languages may pose a challenge when trying to disseminate information to locals about the natural environment and communicate new rules about usage of plastic products. This should to be taken into account when giving out messages to the public. Locals may have difficulty reading information in their non-native language and having information in one language will not be effective in educating all audiences. Therefore, information should be presented in multiple languages to be the most effective at notifying the public. While using various platforms for dissemination whether it is information online, hanging or handing out signs, word of mouth, presentations, and/or implementing this information into the education system, language is a challenge.

These types of dissemination also pose some issues themselves. Some of these types of distribution, for example, putting up signs and information traveling by word of mouth may not be the most reliable ways to get information out to the local public. Furthermore, implementing information into the education system will pose to be a problem because it is difficult to implement new curriculum. Even though it may be difficult, this is an important part of communication so that younger generations can be brought up with more environmentally conscious mind-sets and respect and care for their natural environments. However, there are other types of dissemination that many locals use daily and may be important to use for communication, such as local radio broadcasts.

Enforcing legislations can also be a potential problem from possible insufficient amount of human resources available. The main goal should be to get enforcement department to implement the regulations and if there is no enough manpower, then this may prove to be difficult. The other possible issue is the lack of money to fund the project. Therefore, it is important to have stakeholder involvement. Without sufficient resources, regulating and enforcing a ban may not be effective.

Lastly, suitable alternatives to plastic products need to be provided so that locals and tourists can use them instead of relying on plastic products that have been banned.

REFERENCES

1. Abatneh Yasabie. (2013). Preliminary study on the conversion of different waste plastics into fuel oil. Berlin: GRIN Verlag,.
2. Ackerman Kenneth B. (2013). Practical Handbook of Warehousing. New York: Springer.
3. Andrady, A. L. (2003). Plastics and the Environment. Berlin: John Wiley & Sons.
4. Andrady, A. L. (2015). Plastics and Environmental Sustainability: Fact and Fiction. Berlin: John Wiley & Sons.
5. Anna Frånc, Å. S.-A.-R. (2014). Collection & recycling of plastic waste: Improvements in existing collection and recycling systems in the Nordic countries. London: Nordic Council of Ministers.
6. Anna Fråne, Åsa Stenmarck, Stefán Gíslason, Kari Anne Lyng, Søren Løkke, Malin zu Castell-Rüdenhausen, Margareta Wahlström. (2014). Collection & recycling of plastic waste: Improvements in existing collection and recycling systems in the Nordic countries. Sweden: Nordic Council of Ministers.

7. Bart, J. C. (2006). Polymer Additive Analytics: Industrial Practice and Case Studies. New York: Firenze University Press.
8. Blakistone, B. A. (1999). Principles and Applications of Modified Atmosphere Packaging of Foods. Berlin: Springer Science & Business Media.
9. Boustead, I. (1994). Plastics and the Environment. Chicago: iSmithers Rapra Publishing.
10. D'Mello, J. P. (2003). Food Safety: Contaminants and Toxins. Nairobi: CABI.
11. Edward Denison, G. Y. (2001). Packaging Prototypes 3: Thinking Green. London: RotoVision.
12. Geoff A. Giles, D. R. (2000). Materials and Development of Plastics Packaging for the Consumer Market. London: Taylor & Francis.
13. Geoff A. Giles, D. R. (2001). Technology of Plastics Packaging for the Consumer Market. London: CRC Press.
14. George D. Saravacos, A. E. (2002). Handbook of Food Processing Equipment. Berlin: Springer Science & Business Media.
15. Gleick, P. H. (New York). Bottled and Sold: The Story Behind Our Obsession with Bottled Water. 2010: Island Press.
16. Griffin, R. C. (2012). Principles of Package Development. Berlin: Springer Science & Business Media.
17. H. Lockhart, F. A. (1996). Packaging of Pharmaceuticals and Healthcare Products. Berlin: Springer Science & Business Media.
18. Hannay, F. (2002). Rigid Plastics Packaging: Materials, Processes and Applications. London: iSmithers Rapra Publishing.
19. Horrocks, A. R. (1996). Recycling Textile and Plastic Waste. Chicago: Elsevier.
20. Hulse, S. (2000). Plastics Product Recycling: A Rapra Industry Analysis Report. New York: iSmithers Rapra Publishing.
21. Jakobsen, J. (2003). A Survey of Nordic Packaging Data Collection Methods. Sweden: Nordic Council of Ministers.
22. Jeanne Wirka, J. S. (1989). Wrapped in Plastic: Plastics packaging : the growth of an industry, Part 1. London: Legislative Reference Bureau.

23. John Baxter, M. W.-R. (2015). Plastic value chains: Case: WEEE (Waste Electrical and Electronic Equipment): Part 2 Report. London: Nordic Council of Ministers.
24. John Baxter, Margareta Wahlstrom, Malin Zu Castell-Rüdenhausen, Anna Fråne, Malin Stare, Søren Løkke, Massimo Pizzol. (2014). Plastic value chains: Case: WEEE (Waste Electric and electronic equipment) in the Nordic region. London: Nordic Council of Ministers.
25. José Aguado, D. P. (1999). Feedstock Recycling of Plastic Wastes. London : Royal Society of Chemistry.
26. Knight, G. (2013). Plastic Pollution. California : Raintree.
27. Laws, E. A. (2000). Aquatic Pollution: An Introductory Text. London: John Wiley & Sons.
28. Laws, E. A. (2007). Aquatic Pollution: An Introductory Text. New York: John Wiley & Sons.
29. Magnus Hennlock, Malin zu Castell-Rüdenhausen, Margareta Wahlström, Birgitte Kjær, Leonidas Milios, Eldbjørg Vea, David Watson, Ole Jørgen Hanssen, Anna Fråne, Åsa Stenmarc, Haben Tekie. (2015). Economic Policy Instruments for Plastic Waste: A review with Nordic perspectives. New York: Nordic Council of Ministers.
30. Mantia, F. P. (1996). Recycling of PVC and Mixed Plastic Waste. London: ChemTec Publishing.
31. McKeen, L. W. (2011). Permeability Properties of Plastics and Elastomers. Berlin: William Andrew.
32. McKeen, L. W. (2012). Film Properties of Plastics and Elastomers. Berlin: William Andrew.
33. Morgan, S. (2009). Waste, Recycling and Reuse. London: Evans Brothers.
34. N. G. McCrum, C. P. Buckley, C. B. Bucknall. (1997). Principles of Polymer Engineering. London: Oxford University Press.
35. Niir Board Of Consultants & Engineers. (2000). Medical, Municipal and Plastic Waste Management Handbook. London: National Institute Of Industrial Re.

36. Nottingham, U. o. (2003). Sustainability: The Business Perspective. London: University of Nottingham.
37. Otto G. Piringer, A. L. Baner. (2008). Plastic Packaging Materials for Food: Barrier Function, Mass Transport, Quality Assurance, and Legislation. New York: John Wiley & Sons.
38. Otto G. Piringer, A. L. Baner. (2008). Plastic Packaging: Interactions with Food and Pharmaceuticals. New York: John Wiley & Sons.
39. Paul, M. (2015). One Plastic Bag: Isatou Ceesay and the Recycling Women of the Gambia. Lagos: Millbrook Press.
40. Peter Ulrich, C. S. (2012). Facing Public Interest: The Ethical Challenge to Business Policy and Corporate Communications. Berlin: Springer Science & Business Media.
41. Plastics Instit. (1991). Plastics in Food Packaging Conference. New York: CRC Press.
42. Programme, U. N. (2005). Theo and the Giant Plastic Ball. New York: UNEP/Earthprint.
43. Programme, U. N. (2005). Unep Annual Report 2004. New York: UNEP/Earthprint.
44. Programme, U. N. (2011). Unep 2010 Annual Report. New York : UNEP/Earthprint.
45. R. Beswick, D. J. Dunn. (2002). Plastics in Packaging: Western Europe and North America. London: iSmithers Rapra Publishing.
46. Richard Coles, M. J. (2011). Food and Beverage Packaging Technology. Berlin: John Wiley & Sons.
47. Rien Van den Hondel, Julie Robinson. (2007). European Plastics & Rubber Directory. Berlin: iSmithers Rapra Publishing.
48. Schlosser, E. (2001). Fast Food Nation: The Dark Side of the All-American Meal. Boston: Houghton Mifflin Harcourt.
49. Sidwell, J. (1992). Food Contact Polymeric Materials. London: iSmithers Rapra Publishing.
50. Sina Ebnesajjad. (2012). Plastic Films in Food Packaging: Materials, Technology and Applications. New York: William Andrew.
51. Terry, B. (2015). Plastic-Free: How I Kicked the Plastic Habit and How You Can Too. New York: Skyhorse Publishing.

52. Tukker, A. (2002). Plastics Waste: Feedstock Recycling, Chemical Recycling and Incineration, Volume 13. New York: iSmithers Rapra Publishing.
53. UNEP/Earthprint,. (2011). UNEP Year Book 2011: Emerging Issues in Our Global Environment, Book 2011. New York: UNEP/ Earthprint,.
54. Valuplast Limited. (1997). Plastics Packaging Recycling: One-day Seminar Jointly Organised by Rapra Technology Limited and Valuplast Limited Held on Thursday 20th March 1997 at Rapra Technology Limited, Shawbury, Shrewsbury, Shropshire SY4 4NR, UK. London: iSmithers Rapra Publishing.

INDEX

U

W